T0224006

SpringerBriefs in Applied Sciences and Technology

More information about this series at http://www.springer.com/series/8884

Brajesh Kumar Kaushik · V. Ramesh Kumar
Amalendu Patnaik

Crosstalk in Modern On-Chip Interconnects

A FDTD Approach

 Springer

Brajesh Kumar Kaushik
Department of Electronics and
   Communication Engineering
Indian Institute of Technology Roorkee
Roorkee, Uttarakhand
India

Amalendu Patnaik
Department of Electronics and
   Communication Engineering
Indian Institute of Technology Roorkee
Roorkee, Uttarakhand
India

V. Ramesh Kumar
Department of Electronics and
   Communication Engineering
Indian Institute of Technology Roorkee
Roorkee, Uttarakhand
India

ISSN 2191-530X   ISSN 2191-5318   (electronic)
SpringerBriefs in Applied Sciences and Technology
ISBN 978-981-10-0799-6   ISBN 978-981-10-0800-9   (eBook)
DOI 10.1007/978-981-10-0800-9

Library of Congress Control Number: 2016934012

Printed on acid-free paper

This Springer imprint is published by Springer Nature
The registered company is Springer Science+Business Media Singapore Pte Ltd.

Preface

Advancement in VLSI technology offers gigascale integrated circuits in a system on-chip. In such circuits, interconnects play a key role in determining circuit performance such as time delay and power consumption. At high operating frequencies, the closely packed interconnects produce transient crosstalk. The crosstalk noise strongly influences the signal propagation delay and also causes logic or functional failure. Therefore, it is desirable to accurately model the crosstalk effects in the on-chip interconnects.

Over the years, several mathematical models have been proposed for the analysis of CMOS-gate-driven coupled interconnect lines. However, most of these crosstalk noise models consider the nonlinear CMOS driver as a linear resistor. This approximation is not valid for on-chip interconnects because during the input and output transition states the MOSFET operates in cutoff, linear, and saturation regions. The MOSFET operating time in the saturation region is about 50 % during the transition period. Moreover, the equivalent resistance value in saturation region is much higher than the linear region. Thus, assuming that the transistor operates in the linear region during the transition state leads to severe errors in the performance estimation of the driver interconnect load system. Therefore, it is necessary to develop an accurate model that appropriately considers the nonlinear effects of CMOS driver and accurately measures the crosstalk-induced performance parameters of on-chip interconnects. This book presents an accurate and time efficient model of CMOS-gate-driven coupled interconnects for crosstalk-induced performance analysis by considering the nonlinear effects of CMOS driver.

The conventional interconnect copper material suffers due to lower reliability with downscaling of interconnect dimensions. The reliability of Cu decreases due to electromigration-induced problems such as hillock and void formations. Moreover, with highly scaled dimensions the resistivity of copper increases due to electron-surface scattering and grain-boundary scattering. Therefore, researchers are forced to find an alternative material for on-chip interconnects. Carbon nanotubes (CNTs) have been proposed as a promising interconnect material. A portion of this book is focused toward modeling of MWCNT interconnects. Based on the electrical equivalent model, an accurate FDTD model is presented while incorporating the quantum effects of nanowire and nonlinear effects of CMOS driver. The crosstalk noise is comprehensively analyzed by examining both functional and dynamic crosstalk effects.

Graphene nanoribbon (GNR), a strip of ultrathin width graphene layer, has also been considered aggressively by researchers as a potential alternative material for realizing on-chip interconnects. Most of the physical and electrical properties of GNRs are similar to that of CNTs; however, the major advantage of GNRs over CNTs is that both transistor and interconnect can be fabricated on the same continuous graphene layer, thus avoiding the metal-graphene contact problems. This book presents an accurate model for the analysis of MLGNR interconnects using the FDTD technique. In a more realistic manner, the model incorporates the width-dependent MFP parameter that helps in accurately estimating the crosstalk-induced performance in comparison to conventional models. Moreover, a comparative analysis of crosstalk-induced performance is presented among Cu, MWCNT, and MLGNR interconnects.

The stability of the FDTD technique is constrained by the Courant-Friedrichs-Lewy (CFL) stability condition. Hence, beyond the CFL condition, the technique is unstable and within it, the technique is inefficient. The efficiency improvements in the FDTD technique can be easily addressed if the CFL stability condition is removed by implicitly deriving the transmission line equations. To improve the efficiency of the FDTD technique, an unconditionally stable FDTD (US-FDTD) technique is presented for the analysis of MLGNR interconnects.

This book provides an accurate FDTD model for on-chip interconnects, covering most recent advancements in materials and design. Furthermore, depending on the geometry and physical configurations, different electrical equivalent models for CNT and GNR-based interconnects are presented. Based on the electrical equivalent models the performance comparison among the Cu, CNT, and GNR-based interconnects are also discussed. The proposed models are validated with the HSPICE simulations. The organization of the book is as follows: Chap. 1 introduces the current research scenario in modeling of on-chip interconnects. Chapter 2 presents the structure, properties, and characteristics of graphene based on-chip interconnects. The FDTD modeling of Cu-based on-chip interconnects is presented in Chap. 3. The model considers the nonlinear effects of CMOS driver as well as the transmission line effects of interconnect line that includes coupling capacitance and mutual inductance effects. Chapter 4 introduces an equivalent single conductor (ESC) model of MWCNT interconnects. Based on the ESC model, this chapter presents an accurate FDTD model of MWCNT while incorporating the quantum effects of nanowire and nonlinear effects of CMOS driver. The modeling of MLGNR interconnects using the FDTD technique is presented in Chap. 5. In a more realistic manner, the proposed model includes the effect of width-dependent MFP of the MLGNR while taking into account the edge roughness. Finally, to improve the efficiency of the FDTD model, an unconditionally stable FDTD technique is presented for the analysis of on-chip interconnects in Chap. 6. Moreover, the performance of Cu interconnect is compared with MWCNT and MLGNR interconnects under the influence of crosstalk.

Contents

About the Authors

Brajesh Kumar Kaushik received the B.E. degree in Electronics and Communication Engineering from the D.C.R. University of Science and Technology (*formerly C. R. State College of Engineering*), Murthal, Haryana, in 1994, the M.Tech degree in Engineering Systems from Dayalbagh Educational Institute, Agra, India, in 1997, and the Ph.D. degree under AICTE-QIP scheme from the Indian Institute of Technology Roorkee, Roorkee, India, in 2007. He served at Vinytics Peripherals Pvt. Ltd., Delhi, from 1997 to 1998 as the Research and Development Engineer for microprocessor-, microcontroller-, and DSP processor-based systems.

He joined the department of Electronics and Communication Engineering, G.B. Pant Engineering College, Pauri Garhwal, Uttarakhand, India as a Lecturer in July 1998, where he later served as Assistant Professor from May 2005 to May 2006, and from May 2006 to December 2009. He is currently serving as Associate Professor in the department of Electronics and Communication Engineering, Indian Institute of Technology Roorkee. His research interests include high-speed interconnects, low-power VLSI design, carbon nanotube-based designs, organic thin-film transistor design and modeling, and spintronics-based devices and circuits. He has published extensively in several national and international journals and conferences of repute. Dr. Kaushik is a reviewer of many international journals belonging to various publication houses such as IEEE, IET, Elsevier, Springer, Emerald, Taylor and Francis, etc. He has also delivered many keynote addresses in reputed international and national conferences. He is the editor and editor-in-chief of various journals in the field of VLSI and Microelectronics. Dr. Kaushik is Editor-in-Chief of *International Journal of VLSI Design and Communication System (VLSICS)*, AIRCC Publishing Corporation. He is also the Editor of *Microelectronics Journal (MEJ)*, Elsevier Inc.; *Journal of Engineering, Design and Technology (JEDT)*, Emerald Group Publishing Limited; and *Journal of Electrical and Electronics Engineering Research (JEEER)*, Academic Journals. He is a Senior Member of IEEE and has received many awards and recognitions from the International Biographical Center, Cambridge, U.K. His name has been listed in Marquis Who's Who in Science and Engineering and Marquis Who's Who in the World.

V. Ramesh Kumar received the B.Tech degree in Electronics and Communication Engineering from Bapatla Engineering College, Andhra Pradesh, in 2007, and the M.Tech degree from National Institute of Technology Hamirpur, in 2010. He is currently working toward the Ph.D. degree from Indian Institute of Technology Roorkee, India. His current research interests include time domain numerical methods to approach fast transients characterization techniques, modeling of VLSI on-chip interconnects, carbon based nano-interconnects and through silicon vias.

Amalendu Patnaik received his Ph.D. in Electronics from Berhampur University in 2003. He is currently serving as Associate Professor in the department of Electronics and Communication Engineering, Indian Institute of Technology Roorkee. He served as a Lecturer in National Institute of Science and Technology, Berhampur, India. During 2004–05, he has been to University of New Mexico, Albuquerque, USA as a Visiting Scientist. He has published more than 50 papers in journals and conferences, co-authored one book on Engineering Electromagnetics, and one book chapter on *Neural Network for Antennas in Modern Antenna Handbook* from Wiley. Besides this, he has presented his research work as short courses/tutorials in many national and international conferences. His current research interests include array signal processing, application of soft-computing techniques in Electromagnetics, CAD for patch antennas, EMI, and EMC. He was awarded the IETE Sir J.C. Bose Award in 1998 and BOYSCAST Fellowship in 2004–05 from Department of Science and Technology, Government of India. Dr. Patnaik is a life member of Indian Society for Technical Education (ISTE), Senior Member of IEEE, and IEEE AP-S Region 10 Distinguished Speaker.

About the Book

The book provides accurate FDTD models for on-chip interconnects, covering most recent advancements in materials and design. Furthermore, depending on the geometry and physical configurations, different electrical equivalent models for CNT and GNR based interconnects are presented. Based on the electrical equivalent models the performance comparison among the Cu, CNT, and GNR-based interconnects are also discussed in the book. The proposed models are validated with the HSPICE simulations.

The book introduces the current research scenario in modeling of on-chip interconnects. It presents the structure, properties, and characteristics of graphene-based on-chip interconnects and the FDTD modeling of Cu-based on-chip interconnects. The model considers the nonlinear effects of CMOS driver as well as the transmission line effects of interconnect line that includes coupling capacitance and mutual inductance effects. In a more realistic manner, the proposed model includes the effect of width-dependent MFP of the MLGNR while taking into account the edge roughness.

Chapter 1
Introduction to On-Chip Interconnects and Modeling

Abstract This chapter briefs about the challenges associated with the modeling of on-chip interconnects in nanoscale technology. Copper had been used as an on-chip interconnect material for a long time. However, as device dimensions are scale down the reliability decreases due to electromigration induced problems. Therefore, researchers are forced to find an alternative solution for future high-speed global VLSI interconnects. This chapter introduces the evolution of graphene interconnect materials and the challenges associated with them. This chapter also introduces the FDTD technique for the modeling of on-chip interconnects.

Keywords Carbon nanotube (CNT) · Copper · Finite-difference time-domain (FDTD) · Graphene nanoribbon (GNR) · Interconnects · Very large scale integration (VLSI)

1.1 Introduction

Advancement of technology in the nanometer regime considers high-speed and high-density very large scale integration (VLSI) circuits. It is desirable to use multilayer interconnections in three or more levels to achieve higher packing densities and smaller footprint [1, 2]. Based on the length and cross-sectional dimensions, the on-chip interconnects can be broadly characterized into three categories: local, intermediate, and global interconnects. Local interconnects consist of very thin lines, used to connect gates and transistors in a functional block. Intermediate interconnects are wider and longer than local interconnects, provide low-resistance signal paths in a functional block. The global interconnects provide long-distance communication between the functional blocks and have a large cross-sectional area to minimize the resistance [3]. The global interconnects are placed at the higher level of the chip and can be as long as 1–2 cm in current high-performance integrated circuits [1].

In early days, the operating speed of an integrated circuit was limited by the speed of a logic gate. Interconnects between the gates were considered as ideal conductors, where the signal propagates instantaneously. Therefore, the interconnects had little effect on circuit operation. However, after the introduction of submicron semiconductor devices, the ideal behavior of interconnects no longer remains adequate. In fact, the performance of the chip is primarily determined by the interconnect line rather than the device [4].

At high operating frequencies, the closely packed interconnects produce transient crosstalk [5–7]. The undesired effect created on one line due to a signal transmitted on another line is defined as crosstalk. The crosstalk noise strongly influences the signal propagation delay and causes the circuit malfunction or functional failure. Based on the switching transitions in the coupled lines, crosstalk can be broadly classified into functional and dynamic crosstalks. When the victim line is quiescent, a voltage spike appearing on it due to switching in an adjacent line is referred as the functional crosstalk. Dynamic crosstalk appears when the adjacent lines are simultaneously switching either in-phase or out-phase. A change in logic value and propagation delay can be experienced under functional and dynamic crosstalks, respectively. Moreover, the crosstalk noise causes signal overshoot, undershoot and ringing effects. Therefore, accurate estimation of performance parameters, under the effect of crosstalk, gained importance for the design of high-performance on-chip interconnects.

1.2 Evolution of Interconnect Materials

Aluminum had been used for a long time to form interconnect lines because of its compatibility with silicon. However, as device dimensions scale down the reliability decreases due to increasing current density that may lead to electromigration induced problems [2]. In 1997, IBM announced plans to replace aluminum with copper, a metal with lower resistivity than aluminum [2]. Copper provides high current density (10^7 A/cm^2) leading to the electromigration effect being less significant [8]. Later on, it was realized that even Cu was not able to fulfil the demands of high-speed interconnects due to the following reasons:

(i)   The reliability decreases with down scaling of interconnect dimensions due to increase in current density.
(ii) The resistivity increases at lower dimensions, due to grain-boundary scattering and surface scattering.
(iii) The resistivity increases rapidly due to Joule heating.
(iv) The conductivity reduces at high operating frequencies, due to the skin effect.

Therefore, the copper interconnect material is unable to meet the requirements of future technology needs. The widening gap between the requirements of future on-chip interconnect material and the presently used copper material has compelled researchers and designers to look out for novel material solutions. Graphene based nano interconnects have been proposed as a promising solution for the future on-chip interconnects [9–15]. Encouragingly, graphene nano interconnects demonstrate longer mean free paths (MFPs) in the order of several micrometers, higher current densities in excess of 10^9 A/cm^2, and higher thermal stability than copper. These properties create lots of interest among researchers to use these materials as VLSI interconnects [16, 17]. Graphene nano interconnects can be classified into carbon nanotubes (CNTs) and graphene nanoribbons (GNRs).

1.2.1 Carbon Nanotubes (CNTs)

Carbon nanotubes are single layer of graphene sheets rolled up into cylinders with diameters ranging from 1 to 5 nm. The electron transport in metallic CNTs is ballistic that results in movement of electrons without scattering along the nanotube axis and enables a long MFP in the range of micrometers [18–22]. Contrastingly, the MFP of electrons in Cu is limited to a few nanometers. Due to the large MFP and small diameter, the electrons do not scatter as often in CNTs which results in low resistance. This low resistance ensures that the energy dissipated in CNTs is incredibly small. Thus, the problem of dissipated power density can be properly addressed. Moreover, the 1D structure of CNT offers many electrical properties, such as

(i) High quality CNTs have long MFP ranging from 1 to 5 μm that results in ballistic transport phenomenon.
(ii) The strong sigma bonds are useful for high mechanical strength and pi bonds are useful for high conductivity.
(iii) Higher electron mobility ($\sim 10^5$ cm^2/(Vs)) in comparison to Cu ($\sim 10^3$ cm^2/(Vs)) that results in high drift velocity.
(iv) Larger current densities (10^{10} A/cm^2) in comparison to Cu (10^6 A/cm^2) that results in lower electromigration effect.

Depending on the number of concentrically rolled up graphene sheets, CNTs are categorized as single-walled CNTs (SWCNTs) and multi-walled CNTs (MWCNTs) [23, 24]. SWCNT is a single-layer sheet of graphite rolled up into a cylinder. The primary drawback of SWCNT bundle is the non-controllability of its chirality. The metallic and semiconducting properties of CNTs are primarily dependent on their chirality. Statistically, one-third of the CNTs in a bundle are considered to be conducting (i.e., metallic) while the remaining behaves as semiconductors. Morris [25] observed that the SWCNTs with random chiralities do not show any advantage over the conventional interconnect materials. This problem can be rectified by using MWCNTs that consist of multiple layers of graphene sheets arranged in co-axial configuration with the diameters ranging from 2 nm to several tens of nanometers. Due to the large diameters, the MWCNT shells are conductive even if they are of semiconducting chirality because the energy gap is inversely proportional to the shell diameter. For a semiconducting CNT having a diameter greater than 20 nm, the gap between the conduction band and the Fermi level is observed to be smaller

Table 1.1 Electrical properties of CNTs and comparison with other materials [11]

Semiconducting properties			Metallic properties		
Parameter	Semiconducting CNT	Silicon	Parameter	Metallic CNT	Copper
Bandgap (eV)	0.9/diameter	1.12	Mean free path (nm)	10^3	40
Electron mobility (cm^2/Vs)	20×10^3	1500	Current density (A/cm^2)	10^{10}	10^6
Electron phonon mean free path (μm)	~ 0.07	0.0076	Resistivity (Ω-m)	$\sim 10^{-5}$	1.68×10^{-8}

than 0.0258 eV, which can be smeared by the environmental temperature [26]. Table 1.1 summarizes the unique electrical properties of metallic and semiconducting CNTs and compares them with copper and silicon.

1.2.2 Graphene Nanoribbons (GNRs)

Graphene nanoribbon, a strip of ultra-thin width graphene layer, has also been considered aggressively by the researchers as a potential alternative material for realizing on-chip interconnects [27, 28]. Most of the physical and electrical properties of GNRs are similar to that of CNTs, however, the major advantage of GNRs over CNTs is that both transistors and interconnects can be fabricated on the same graphene layer [29]. Therefore, one of the manufacturing difficulties regarding the formation of metal–nanotube contact can be avoided. Depending on the number of stacked graphene sheets, GNRs are classified as single-layer GNRs (SLGNRs) and multilayer GNRs (MLGNRs). Due to the lower resistivity and easy fabrication process, the MLGNRs are often preferred over SLGNRs as interconnect material. However, the MLGNRs fabricated till date have displayed some level of edge roughness [30, 31]. The electron scattering at the rough edges, reduces the MFP that substantially lowers the conductance of the MLGNRs. This fundamental challenge limits the performance of MLGNR interconnects.

1.3 Modeling of On-Chip Interconnects

Historically, interconnects were modeled as a lumped capacitor [1]. With the advancement of technology, the cross-sectional area of interconnects were scaled down, due to which the line resistance became significant and therefore, the interconnect line was represented as lumped resistance-capacitance (RC) [32]. However, later these interconnect parasitic elements were not treated as lumped elements. To improve the accuracy, a distributed *RC* model was considered [33]. Currently, the parasitic inductance has started to play an important role in an on-chip interconnect performance due to the adoption of low resistive interconnect materials, and high operating switching frequencies. Therefore, the on-chip interconnects must be treated as distributed *RLC* lines or as transmission lines to estimate the performance accurately [34]. Agarwal et al. [4] proposed a model considering the transmission line effects of coupled on-chip interconnects driven by a linear resistor. Kaushik et al. extended this model to a nonlinear complementary metal–oxide–semiconductor (CMOS) driver using alpha-power law model and analyzed functional crosstalk effects in [35] and dynamic crosstalk effects in [36]. The models reported in [4, 35] and [36] are based on even-odd modes and hence limited to purely two coupled interconnect lines. Furthermore, the transient analysis was carried out for lossless lines, which is impractical.

The modeling of distributed *RLC* lines along with nonlinear CMOS driver suffers from frequency/time domain conversion problem. This problem arises because the transmission lines were traditionally solved in the frequency domain by using the partial differential equations, whereas the CMOS driver is modeled in the time domain. Therefore, to avoid this conversion problem, most of the researchers [4, 5, 37] replace the nonlinear CMOS driver by the linear resistive driver that severely affects the accuracy of the model. In the present research work, to avoid the conversion problem, finite-difference time-domain (FDTD) technique is used to solve the transmission line equations in the time domain. Using the FDTD method, the voltage and current values can be correctly estimated at any particular point on the interconnect line. Moreover, the FDTD model can be extended to n coupled interconnect lines with low computational cost.

In past, FDTD techniques were used to analyze the transmission lines, which are excited and terminated by resistive driver and resistive load, respectively [38–40]. Including the frequency dependent losses, Orlandi et al. [41] proposed the FDTD model for the analysis of multiconductor transmission lines terminated in arbitrary loads using the state-variable formulation. However, the models proposed in [38–40] analyze the transmission lines with resistive drivers and hence are not valid for the accurate study of on-chip interconnects performance, which are actually excited and terminated by the CMOS inverters. Based on the FDTD technique, Li et al. [42] proposed a model for the transient analysis of CMOS gate-driven distributed *RLC* interconnects. Coupled interconnects were analyzed at global interconnect length using 180 nm technology node where the nonlinear CMOS drivers were modeled by the alpha-power law model. This model is not accurate under the conditions when the technology is scaled down beyond 180 nm, due to the ignorance of the finite drain conductance parameter. Therefore, it is necessary to develop an accurate model that appropriately considers the nonlinear effects of CMOS drivers and accurately measures the crosstalk induced performance parameters of on-chip interconnects. This book presents an accurate and time-efficient model of CMOS gate-driven coupled interconnects for crosstalk induced performance analysis. The model is developed using the FDTD technique for coupled on-chip interconnects, whereas the

CMOS driver is modeled by either nth power law or modified alpha-power law model by considering the finite drain conductance parameter. The model is validated by comparing the results with HSPICE simulations.

1.4   Introduction to FDTD Method

FDTD is a popular computational electromagnetic modeling technique. Initially, this method was implemented to solve Maxwell's equations in the time domain. In this method, the time-dependent partial differential equations are discretized in space and time using the central difference approximations. The resulting finite difference equations are solved in a leapfrog time stepping manner for the solution of the Maxwell's equations [43]. In the present book, FDTD technique is used to solve transmission line equations, where the relative parameters are line voltage (V) and current (I) on a transmission line, that are analogous to the electrical field (E) and magnetic field (H) in Maxwell's equations, respectively.

To carry out the FDTD analysis of a transmission line model, a computational domain has to be established. The computational domain for the transmission line is modeled by resistance, inductance and capacitance (RLC) elements with relative parameters of V and I. The V and I values are determined at each point in space and time within the computational domain. The resistance (R), inductance (L), and capacitance (C) parasitics of the line must be specified for each cell in the computational domain. In case of coupled line system, the mutual inductance and the coupling capacitance should also be included within the computational domain. After establishing the computational domain, the boundary conditions have to be specified at the near-end and far-end boundaries. At the near-end boundary, the interconnect line is driven by a nonlinear CMOS inverter, whereas at the far-end boundary the interconnect line is terminated by a gate capacitance of the CMOS inverter load. After specifying the near and far-end terminals, the boundary conditions must be incorporated in the computational domain to match the FDTD solution for voltage and current at discrete time and space points. Time is implicit in the FDTD method, whereas space is explicit. Nevertheless, the FDTD method provides the exact solution if the following two conditions are satisfied: (1) spatial step (Δz) must be small enough in comparison to the wavelength (generally 10–20 steps per wavelength); (2) time step (Δt) must be small enough to satisfy the Courant condition [38–41]. The FDTD technique offers the following advantages:

(i)   This technique allows the user to specify the computational domain. A wide variety of linear and nonlinear mediums can be modeled easily.
(ii) The voltage and current parameters are evaluated at any particular point on the interconnect line can be quickly and accurately obtained.
(iii) The technique requires less number of assumptions; therefore, the accuracy is very high.
(iv) The technique is able to simulate a wide range of frequencies.
(v)   The model can be extended to n number of lines with low computational cost.

1.4.1 Central Difference Approximation

The FDTD method is developed based on the central difference approximation. This subsection describes the evaluation of numerical derivatives using central difference approximation. Consider a function of one variable $f(t)$. Expanding this in a Taylor series in a neighborhood of a desired point t_0 gives

$$f(t_0 + \Delta t) = f(t_0) + \Delta t f'(t_0) + \frac{\Delta t^2}{2!} f''(t_0) + \frac{\Delta t^3}{3!} f'''(t_0)........ \qquad (1.1)$$

where the primes denote the various derivatives with respect to the function. Solving this for the first derivative gives

$$f'(t_0) = \frac{f(t_0 + \Delta t) - f(t_0)}{\Delta t} - \frac{\Delta t}{2!} f''(t_0) - \frac{\Delta t^2}{3!} f'''(t_0)........ \qquad (1.2)$$

Thus first order derivative can be approximated as

$$f'(t_0) = \frac{f(t_0 + \Delta t) - f(t_0)}{\Delta t} + \theta(\Delta t) \qquad (1.3)$$

where $\theta(\Delta t)$ denotes that the error in truncating the series is on the order of Δt. So the first derivative may be approximated with the forward difference

$$f'(t_0) \cong \frac{f(t_0 + \Delta t) - f(t_0)}{\Delta t} \qquad (1.4)$$

This amounts to approximating the derivative of $f(t)$ as with its region of the desired point. If we expand the Taylor's series as

$$f(t_0 - \Delta t) = f(t_0) - \Delta t f'(t_0) + \frac{\Delta t^2}{2!} f''(t_0) - \frac{\Delta t^3}{3!} f'''(t_0)........ \qquad (1.5)$$

which can be approximated by

$$f'(t_0) \cong \frac{f(t_0) - f(t_0 - \Delta t)}{\Delta t} \qquad (1.6)$$

This gives the backward approximate equation for derivative.

Other approximations known as central differences can be found by subtracting Eq. (3.3) from Eq. (3.6) to yield the first derivative central difference approximation

$$f'(t_0) \cong \frac{f(t_0 + \Delta t) - f(t_0 - \Delta t)}{2\Delta t} \qquad (1.7)$$

with a truncation error of Δt^2. Similarly, the second derivative central difference is obtained by adding (3.2) and (3.6) to yield

$$f''(t_0) \cong \frac{f(t_0 + \Delta t) - 2f(t_0) + f(t_0 - \Delta t)}{\Delta t^2} \qquad (1.8)$$

with a truncation error on the order of Δt^2. Due to second-order error truncation in central difference approximation, FDTD termed as FDTD solution with second-order accuracy.

References

1. Rabaey JM, Chandrakasan A, Nikolic B (2003) Digital integrated circuits: a design perspective, 2nd ed. Prentice-Hall
2. Goel AK (2007) High-speed VLSI interconnections, 2nd ed. Wiley-IEEE Press
3. Rabindra K, Srivastava P, Sharma GK (2010) Network-on-chip: on-chip communication solution. Int Rev Comput Softw 5(1):22–33
4. Agarwal K, Sylvester D, Blaauw D (2006) Modeling and analysis of crosstalk noise in coupled RLC interconnects. IEEE Trans Comput Aided Des Integr Circuits Syst 25(5):892–901
5. Sahoo M, Ghosal P, Rahaman H (2015) Modeling and analysis of crosstalk induced effects in multiwalled carbon nanotube bundle interconnects: an ABCD parameter-based approach. IEEE Trans Nanotechnol 14(2):259–274
6. Kumar VR, Majumder MK, Kaushik BK (2014) Graphene based on-chip interconnects and TSVs—prospects and challenges. IEEE Nanatechnol Mag 8(4):14–20
7. Zhang J, Friedman EG (2006) Crosstalk modeling for coupled *RLC* interconnects with application to shield insertion. IEEE Trans VLSI Syst 14(6):641–646
8. Kaushik BK, Majumder MK, Kumar VR (2014) Carbon nanotube based 3-D interconnects—a reality or a distant dream. IEEE Circuits Syst Mag 14(4):16–35
9. Srivastava A, Xu Y, Sharma AK (2010) Carbon nanotubes for next generation very large scale integration interconnects. J Nanophotonics 4(1):1–26
10. Javey A, Kong J (2009) Carbon nanotube electronics. Springer, Berlin
11. Xu Y (2011) Carbon nanotube interconnect modeling for very large scale integrated circuits. PhD Dissertation, Louisiana State University, USA
12. Das D, Rahaman H (2011) Analysis of crosstalk in single- and multiwall carbon nanotube interconnects and its impact on gate oxide reliability. IEEE Trans Nanotechnol 10(6):1362–1370
13. Tamburrano A, D'Aloia AG, Sarto MS (2012) Bundles of multiwall carbon nanotube interconnects: RF crosstalk analysis by equivalent circuits. In: Proceedings of the IEEE international symposium on electromagnetic compatibility (EMC), Pittsburgh, pp 434–439
14. Plombon JJ (2007) High-frequency electrical properties of individual and bundled carbon nanotubes. Appl Phys Lett 90(6):063106-1–063106-3
15. Xu Y, Srivastava A (2009) A model for carbon nanotube interconnects. Int J Circuit Theory Appl 38(6):559–575
16. Berber S, Kwon YK, Tomanek D (2000) Unusually high thermal conductivity of carbon nanotubes. Phys Rev Lett 84(20):4613–4616
17. Areshkin DA, Gunlycke D, White CT (2007) Ballistic transport in graphene nanostrips in the presence of disorder: importance of edge effects. Nano Lett 7(1):204–210
18. Avorious P, Chen Z, Perebeions V (2007) Carbon-based electronics. Nat Nanotechnol 2 (10):605–13

19. Close GF, Wong HSP (2008) Assembly and electrical characterization of multiwall carbon nanotube interconnects. IEEE Trans Nanotechnol 7(5):596–600
20. Collins PG, Arnold MS, Avouris Ph (2001) Engineering carbon nanotubes and nanotube circuits using electrical breakdown. Science 292(5517):706–709
21. Liang F, Wang G, Lin H (2012) Modeling of crosstalk effects in multiwall carbon nanotube interconnects. IEEE Trans Electromagn Compat 54(1):133–139
22. Narasimhamurthy KC, Paily R (2009) Impact of bias voltage on inductance of carbon nanotube interconnects. In: Proceedings of the 22nd international conference on VLSI design, New Delhi, pp 505–510
23. Majumder MK, Pandya ND, Kaushik BK, Manhas SK (2012) Analysis of MWCNT and bundled SWCNT Interconnects: impact on crosstalk and area. IEEE Electron Device Lett 33 (8):1180–1182
24. Majumder MK, Kaushik BK, Manhas SK (2014) Analysis of delay and dynamic crosstalk in bundled carbon nanotube interconnects. IEEE Trans Electromagn Compat 56(6):1666–1673
25. Morris JE (2008) The proof is in the nanopacking. IEEE Nanotechnol Mag 2(4):25–27
26. Li H, Yin WY, Banerjee K, Mao JF (2008) Circuit modeling and performance analysis of multi-walled carbon nanotube interconnects. IEEE Trans Electron Devices 55(6):1328–1337
27. Murali KH, Brenner K, Yang Y, Beck T, Meindl JD (2009) Resistivity of graphene nanoribbon interconnects. IEEE Electron Device Lett 30(6):611–613
28. Tanachutiwat S, Liu S, Geer R, Wang W (2009) Monolithic graphene nanoribbon electronics for interconnect performance improvement. In: Proceedings of the IEEE international symposium on circuits and systems. Taipei, pp. 589–592
29. Naeemi A, Meindl JD (2008) Electron transport modeling for junctions of zigzag and armchair graphene nanoribbons (GNRs). IEEE Electron Device Lett 29(5):497–499
30. Avouris P (2010) Graphene: electronic and photonic properties and devices. Nano Lett 10 (11):4285–4294
31. Berger C, Song Z, Li X, Wu X, Brown N, Naud C, Mayou D, Li T, Hass J, Marchenkov AN, Conrad EH, First PN, Heer WA (2006) Electronic confinement and coherence in patterned epitaxial graphene. Science 312(5777):1191–1196
32. Bothra S, Rogers B, Kellam M, Osburn CM (1993) Analysis of the effects of scaling on interconnect delay in ULSI circuits. IEEE Trans Electron Devices 40(3):591–597
33. Bakoglu HB (1990) Circuits, interconnections and packaging for VLSI. Addison-Wesley Publishing Company, Boston
34. Ismail YI, Friedman EG (2000) Effects of inductance on the propagation delay and repeater insertion in VLSI circuits. IEEE Trans Very Large Scale Integr VLSI Syst 8(2):195–206
35. Kaushik BK, Sarkar S (2008) Crosstalk analysis for a CMOS-gate-driven coupled interconnects. IEEE Trans Comput Aided Des Integr Circuits Syst 27(6):1150–1154
36. Kaushik BK, Sarkar S, Agarwal RP, Joshi RC (2010) An analytical approach to dynamic crosstalk in coupled interconnects. Microelectron J 41(2):85–92
37. Cui J, Zhao W, Yin W, Hu J (2012) Signal transmission analysis of multilayer graphene nano-ribbon (MLGNR) interconnects. IEEE Trans Electromagn Compat 54(1):126–132
38. Paul CR (2008) Analysis of multiconductor transmission lines. IEEE Press
39. Paul CR (1994) Incorporation of terminal constraints in the FDTD analysis of transmission lines. IEEE Trans Electromagn Compat 36(2):85–91
40. Paul CR (1996) Decoupling the multi conductor transmission line equations. IEEE Trans Microw Theory Tech 44(8):1429–1440
41. Orlandi A, Paul CR (1996) FDTD analysis of lossy, multiconductor transmission lines terminated in arbitrary loads. IEEE Trans Electromagn Compat 38(3):388–399
42. Li XC, Ma JF, Swaminathan M (2011) Transient analysis of CMOS gate driven RLGC interconnects based on FDTD. IEEE Trans Comput Aided Des Integr Circuits Syst 30(4):574–583
43. Livshits P, Sofer S (2012) Aggravated electromigration of copper interconnection lines in ULSI devices due to crosstalk noise. IEEE Trans Device Mater Reliab 12(2):341–346

Chapter 2
Interconnect Modeling, CNT and GNR Structures, Properties, and Characteristics

Abstract This chapter reviews the Cu-based on-chip interconnect modeling. The unique atomic structure and properties of carbon nanotube (CNT) and graphene nanoribbon (GNR) are discussed. The characteristics and semiconducting/metallic properties of graphene-based on-chip interconnects are presented. Depending on the physical configuration, equivalent electrical models of MWCNT and MLGNR interconnect lines are also introduced. An extensive review on performance analysis of on-chip interconnects is presented.

Keywords Electrical equivalent model · Equivalent single conductor (ESC) · Lüttinger liquid theory · Multiconductor transmission line (MTL) · Tight-binding approximation

2.1 Interconnect Modeling Approaches

In the early days of VLSI design, the crosstalk-induced signal integrity effects were negligible because of relatively low integration density and slow operating speed. However, with the introduction of technology scaling of below 0.25 μm, there were many significant changes in the structure and electrical characteristics [1, 2]. The interconnect lines started to be a dominating factor for chip performance and robustness. The line parasitic elements have a major impact on the electrical behavior of the interconnect model. These models vary from simple to very complex depending upon the effects that are being studied and the required accuracy. There are three different types of approaches available in literature for modeling Cu-based on-chip interconnects.

2.1.1 Lumped Model with CMOS Driver

This approach focuses on the CMOS gate modeling while the interconnect line is approximately considered as a lumped circuit. Alpha-power law model [3] has been

© The Author(s) 2016
B.K. Kaushik et al., *Crosstalk in Modern On-Chip Interconnects*,
SpringerBriefs in Applied Sciences and Technology,
DOI 10.1007/978-981-10-0800-9_2

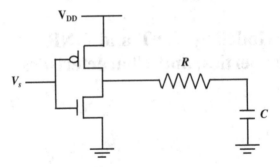

Fig. 2.1 A CMOS gate-driven RC load

widely used for representation of short-channel transistor that includes the velocity saturation effects. Based on the alpha-power law model, delay formulas were developed for CMOS gate-driven lumped capacitance modeled interconnect line [4, 5]. Bisdounis et al. [6] extended the model to include the influences of short-circuit current and gate-to-drain coupling capacitance. With the resistive component of interconnect becoming comparable to the gate output impedance, the line resistance R of interconnect line needs to be considered. Considering the line resistance, the modeling of CMOS gate-driven resistor-capacitor (RC) line was presented in [7–9]. The representation of CMOS gate-driven lumped RC load is shown in Fig. 2.1. Alder and Friedman [10] derived the delay equations for repeater insertion of a CMOS buffer design with RC interconnects. They developed a closed-form expression for the timing analysis of CMOS gate-driven RC load. They also derived an expression for the short-circuit power dissipation of the driver-interconnect-load system. However, all these approaches [3–10], lumped the total wire resistance of each segment into one single R and similarly combined the global capacitance into a single capacitor C. This lumped RC model is inaccurate for long interconnect wires, which are more adequately represented by a distributed RC model. Moreover, due to high operating frequencies and wider interconnect dimensions, interconnects exhibit inductance effects and should be included in the delay and crosstalk noise models. Hence, the analytical models that considered only RC were no longer accurate [3–10].

2.1.2 Distributed Model with Resistive Driver

In the distributed model with resistive driver, the driver-interconnect-load system is analyzed by simplifying the CMOS gate driver as a resistive driver [11]. Using the linear driver approximation, Elmore delay model was initially developed for RC lines [12] and then extended to RLC lines [13, 14]. The distributed RLC line with linear driver is shown in Fig. 2.2, where R_d and C_L represent driver resistance and load capacitance, respectively; r, l and c represent per-unit-length line resistance,

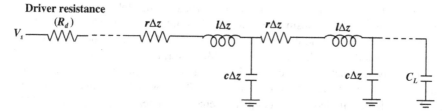

Fig. 2.2 A distributed *RLC* interconnect line driven by a resistive driver

inductance, and capacitance, respectively. Considering the nonlinearity of the driver, Bai et al. [15] improved the linear driver model by calculating the effective resistance. Davis and Meindl proposed closed-form delay expressions for the analysis of distributed *RLC* lines by considering the single transmission line effects in [16] and a crosstalk noise model of coupled transmission lines in [17].

Based on even–odd mode technique, the crosstalk noise model was developed by Agarwal et al. [18] for coupled-two lossless lines and then modified for low-loss lines to analyze crosstalk-induced noise peak voltage. They investigated that at high operating frequencies, inductive coupling effects are significant and should be included for accurate crosstalk noise analysis. Using the coupled transmission line theory, the authors developed a crosstalk noise model that is useful to guide noise-aware physical design optimizations. A closed-form analytical transient response model was derived for resistance/capacitance loads by solving semifinite transmission line equations [19]. However, all these models [11–19] consider the nonlinear CMOS driver as a linear driver that limits the accuracy of the models.

2.1.3 Distributed Model with CMOS Driver

The distributed model with CMOS driver approach co-simulates the nonlinear CMOS gate and the distributed interconnect. The CMOS gate-driven distributed *RLC* interconnect line is shown in Fig. 2.3. Based on the even–odd modes, Kaushik et al. proposed a simple analytical model for functional crosstalk analysis in [20] and dynamic crosstalk analysis in [21] of CMOS gate-driven two coupled

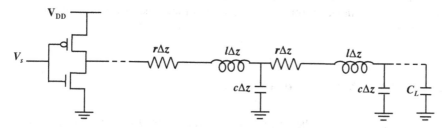

Fig. 2.3 A CMOS gate-driven distributed *RLC* interconnect line

interconnect lines. The model was developed based on the alpha-power law model of MOS transistors and transmission line theory of interconnects. There the authors have observed that the nonlinear effects of the CMOS inverter should be incorporated in a valid crosstalk noise modeling. Moreover, it was noticed that the use of resistive driver model presented a pessimistic view on the performance analysis of on-chip interconnects. However, these models [20, 21] were based on the even–odd modes and hence limited to two coupled interconnect lines. Later on, Li et al. [22] proposed an FDTD method for the transient analysis of CMOS gate-driven lossy transmission lines including frequency-dependent losses and observed the effect of functional crosstalk. However, the model ignored the finite drain conductance parameter in the modeling of CMOS driver and hence not useful for nanoscaled devices.

2.2   Carbon Nanotubes

Until mid-1980s, diamond and graphite were the only two known forms of carbon allotropes. In 1985, Kroto et al. [23] were able to synthesize new allotrope of carbon C_{60}. They used a high pulse of laser light to vaporize a sample of graphite. The vaporized graphite was sent to a mass spectrometer with the help of helium gas. The mass spectrometer detected the presence of C_{60}, a molecule consisting of 60 carbon atoms. The C_{60} had the shape of a soccer ball with 12 pentagon faces and 20 hexagonal faces. The easy synthesis of C_{60} led the group to propose the existence of another allotrope of carbon named as "buckyball" due to its soccer ball-shaped structure. The shape of the new allotrope of carbon did not end at the soccer-shaped structures and long cylindrical tube-like structures were also reported, which are known as carbon nanotubes (CNTs).

CNTs have fascinated the research world due to their extraordinary physical, electrical, and chemical properties. Many of the properties defy the conventional trends and scientists are still discovering the unique properties and constantly making efforts to understand and explain the phenomenon for such distinctive behavior. One of the remarkable physical properties of CNT is its ability to scale down its thickness to a single atomic layer. Another interesting physical property observed in CNTs is when two slightly different-structured CNTs are joined together; the resultant junction formed can be used as an electronic device. The properties of the device formed are dependent on the type of CNTs used for their formation.

2.2.1   Basic Structure of CNTs

A single-walled CNT can be assumed as a structure formed when a single graphene sheet is rolled into a cylindrical shape (Fig. 2.4). Depending on the shape of the

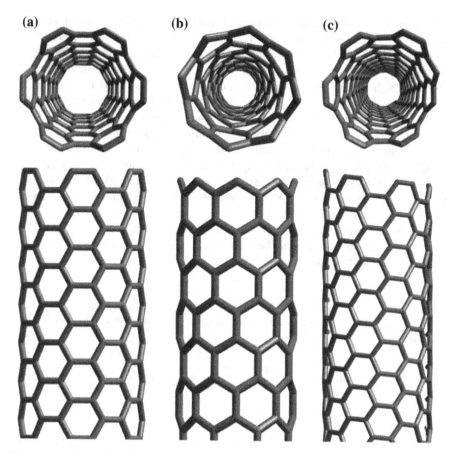

(a) **(b)** **(c)**

Fig. 2.4 Sketches of three different SWCNT structures: **a** armchair nanotube, **b** zigzag nanotube, and **c** chiral nanotube

circumference, CNTs can be classified as armchair (ac), zigzag (zz), or chiral CNTs as shown in Fig. 2.4a–c, respectively [24]. The terms zigzag and armchair are inspired from the pattern in which the carbon atoms are arranged at the edge of the nanotube cross section. Graphene consists of sp^2-hybridized atoms of carbon that are arranged in a hexagonal pattern. The hexagonal carbon rings should join coherently when placed in contact with adjacent carbon atoms. Accordingly, in an SWCNT tube, all the carbon atoms (except at the edges) form hexagonal rings and are therefore equally spaced from one another. Xu et al. [25] reported the fabrication of vertically grown CNT bundles with an average diameter of 50 μm and a pitch of 110 μm.

In spite of the hexagonal aromatic rings, SWCNT is considered to be more reactive than planar graphene. It is due to the fact that the hybridization in SWCNTs is not purely sp^2 and some degree of sp^3 hybridization is also present. It has been observed that with the decrease in SWCNT diameter, the degree of sp^3

hybridization increases [26]. This phenomenon causes variable overlapping of energy bands that results in SWCNTs obtaining versatile and unique electrical properties. It was studied that beyond the diameter of ≈2.5 nm, the SWCNT tube collapses into a two-layer ribbon [26]. Moreover, a CNT with smaller diameter results in higher stress on the structure, although SWCNTs of ≈0.4 nm diameter have been produced [27]. It is therefore natural to consider that a diameter of ≈1 nm is the most suitable value with regard to energy consideration of SWCNTs. Encouragingly, there are no such restrictions on the length of the SWCNTs. The length is dependent on the processes and methods used for synthesis of the SWCNTs. SWCNTs of length ranging from micrometers to millimeters can be commonly observed. Considering the diameter and length of an SWCNT, it is easy to intuitively conclude that SWCNT structures have exceptionally high aspect ratios.

A graphene sheet can be rolled in a number of different ways (see Fig. 2.4). The mathematical expression that can be used to represent the various ways of rolling graphene into the tubes is shown below [28]:

$$PX = C_h = pb_1 + qb_2 \tag{2.1}$$

where C_h is chirality vector, p and q are integers. The unit vectors b_1 and b_2 are defined as

$$b_1 = \frac{b\sqrt{3}}{2}x + \frac{b}{2}y \quad \text{and} \quad b_2 = \frac{b\sqrt{3}}{2}x - \frac{b}{2}y \tag{2.2}$$

where $b = 2.46$ Å and $\cos \theta = \frac{2p+q}{2\sqrt{p^2+q^2+pq}}$.

The vector PX is normal to the CNT tube axis and θ is chirality angle. The diameter d of a nanotube is dependent on C_h by the following relation:

$$d = \frac{|C_h|}{\pi} = \frac{b_{C-C}\sqrt{3(p^2+q^2+pq)}}{\pi} \tag{2.3}$$

where 1.41 Å(graphene) $\leq b_{C-C} \leq 1.44$ Å(C_{60}).

The C–C bond length in the hexagonal ring structure of SWCNT slightly increases from the C–C bond length in graphene due to the curved structure of SWCNT. The degree of curvature in an SWCNT cannot exceed the degree of curvature in C_{60} molecule, resulting in the upper limit of C–C bond length in SWCNTs. Similarly, the degree of curvature in an SWCNT cannot be less than the curvature in a flat graphene structure, resulting in lower limit of C–C bond length in SWCNTs. Moreover, it can be observed that C_h, θ, and d can be expressed in terms of p and q. Since SWCNTs can be identified by C_h, θ, and d values, it is sufficient to define SWCNTs through p, q values by denoting them as (p, q). The p and q values for a particular SWCNT can be easily obtained by counting the number of hexagonal rings separating the margins of C_h vector following b_1 first and then b_2

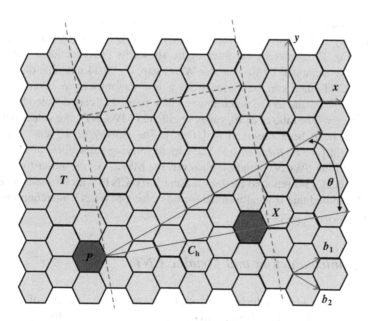

Fig. 2.5 Sketch representing the procedure to obtain a CNT, starting from a graphene sheet

[29]. Based on the (p, q) representation, zz SWCNTs can be denoted as $(p, 0)$ and having $\theta = 0°$; ac SWCNTs can be denoted as (p, p) and having $\theta = 30°$; chiral SWCNTs can be denoted as (p, q) and having $0 < \theta < 30°$.

From Fig. 2.5, it can be observed that having C_h direction perpendicular to any carbon bond directions results in zz SWCNT ($\theta = 0°$), while having C_h direction parallel to any carbon bond directions will result in ac SWCNT ($\theta = 30°$). In chiral SWCNTs, $0 < \theta < 30°$ due to hexagonal rings in graphene sheet.

An MWCNT is a bit more intricate in structure compared to an SWCNT. Unlike a single graphene shell in an SWCNT, there are multiple graphene shells in an

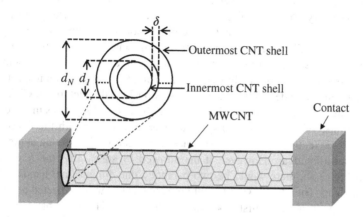

Fig. 2.6 The structure of MWCNT placed between the two contacts

MWCNT. The MWCNTs have two or more numbers of CNT shells that are concentrically rolled up. The structure of MWCNT between the two contacts is shown in Fig. 2.6, wherein the inset figure shows its cross-sectional view. The intershells are separated by the van der Waals gap, $\delta \sim 0.34$ nm. The diameter of outermost CNT shell can be varied from a few nanometers to several tens of nanometers. The diameters of outermost and innermost shells are denoted by d_N and d_1, respectively. The ratio of d_1/d_N varies in different MWCNTs, the values between 0.3 and 0.8 have been observed in [30–32]. The density of $10^6/cm^2$ has been obtained in [30] with a d_1/d_N of 0.5. The current carrying capabilities of MWCNTs are similar to the SWCNT bundles, however, the MWCNTs are easier to fabricate [33]. Close et al. [34] reported the fabrication of MWCNTs with 80 shells based on a versatile method that is ideally suited for fabricating MWCNT interconnects with extensive electrical properties.

2.2.2 Semiconducting and Metallic CNTs

CNTs can act as semiconducting or metallic based on the pattern of CNT circumference. The armchair CNTs always act as metallic, whereas the zigzag CNTs act as either metallic or semiconducting depending on the chiral indices. This section presents the behavior of zz CNTs and their dual nature.

Since CNT is a rolled-up sheet of graphene, an appropriate boundary condition is required to explore the band structure. If CNT can be considered as an infinitely long cylinder, there are two wave vectors associated with it: (1) the wave vector parallel to CNT axis, $k_{\parallel}$, that is continuous in nature due to the infinitely long length of CNTs and (2) the perpendicular wave vector, $k_{\perp}$, that is along the circumference of the CNT. These two wave vectors must satisfy a periodic boundary condition [28]

$$k_{\perp} . C_h = \pi d k_{\perp} = 2\pi m \qquad (2.4a)$$

where m is an integer. The quantized values of allowed $k_{\perp}$ for CNTs are obtained from the boundary condition. The cross-sectional cutting of the energy dispersion with the allowed $k_{\perp}$ states results in the one-dimensional band structure of graphene as shown in Fig. 2.7a. This is called zone folding scheme of obtaining the band structure of CNTs. Each cross-sectional cutting gives rise to one-dimensional subband. The spacing between allowed $k_{\perp}$ states and their angles with respect to the surface Brillouin zone determine the one-dimensional band structures of CNTs. The band structure near the Fermi level is determined by allowed $k_{\perp}$ states that are close to the K points. When the allowed $k_{\perp}$ states pass directly through the K points as shown in Fig. 2.7c, the energy dispersion has two linear bands crossing at the Fermi level without a bandgap. However, if the allowed $k_{\perp}$ states miss the K points as shown in Fig. 2.7b, there would be two parabolic one-dimensional bands with an energy bandgap. Therefore, two different kinds of CNTs can be expected depending on the wrapping indices, first, the semiconducting CNTs with bandgap as in Fig. 2.7b and second, the metallic CNTs without bandgap as in Fig. 2.7c [28].

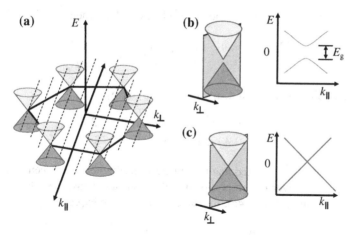

Fig. 2.7 Band structures of CNT shell **a** first Brillouin zone of graphene with conic energy dispersions at six K points. The allowed $k_\perp$ states in CNT are presented by dashed lines. The band structure of the CNT is obtained by the cross sections as indicated. Close view of the energy dispersion near one of the K points are schematically shown along with the cross sections by allowed $k_\perp$ states and resulting one-dimensional energy dispersions for **b** a semiconducting CNT and **c** a metallic CNT

Using the approach of one-dimensional subbands discussed in previous subsection, the one-dimensional subband closest to the K points for zigzag CNTs is investigated here. Based on the chiral indices, the zigzag CNTs can show metallic/semiconducting property. Since the circumference is $\bar{p} \cdot b (C_h = \bar{p} \cdot b_1)$, the boundary condition in Eq. (2.4a) becomes

$$k_x \bar{p} b = 2\pi m \qquad (2.4b)$$

There is an allowed k_x that coincides with K point at $(0, 4\pi/3b)$. This condition arises when $\bar{p}$ has a value in multiples of 3 ($\bar{p} = 3\bar{q}$, where $\bar{q}$ is an integer). Therefore, by substitution in Eq. (2.4b) [28]

$$k_x = \frac{2\pi m}{\bar{p} b} = \frac{3Km}{2\bar{p}} = \frac{Km}{2\bar{q}} \qquad (2.5)$$

There is always an integer ($m = 2\bar{q}$) that makes k_x pass through K points and these types of CNTs (with $\bar{p} = 3\bar{q}$) are always metallic without bandgap as shown in Fig. 2.7c. There are two cases when p is not a multiple of 3. If $\bar{p} = 3\bar{q} + 1$, the k_x is closest to the K point at $m = 2\bar{q} + 1$ (as in Fig. 2.7b).

$$k_x = \frac{2\pi m}{\bar{p} b} = \frac{3Km}{2\bar{p}} = \frac{3K(2\bar{q} - 1)}{2(3\bar{q} + 1)} = K + \frac{K}{2} \frac{1}{3q + 1} \qquad (2.6)$$

Similarly, for $\bar{p} = 3\bar{q} - 1$, the allowed k_x closest to K is when $m = 2\bar{q} - '1$, hence

$$k_x = \frac{2\pi m}{\bar{p}b} = \frac{3Km}{2\bar{p}} = \frac{3K(2\bar{q} - 1)}{2(3\bar{q} - 1)} = K - \frac{K}{2}\frac{1}{3\bar{q} - 1} \tag{2.7}$$

In these two cases, allowed k_x misses the K point by

$$\Delta k_x = \frac{K}{2}\frac{1}{3\bar{q} \pm 1} = \frac{2}{3}\frac{\pi}{\bar{p}b} = \frac{2}{3}\frac{\pi}{\pi d} = \frac{2}{3d} \tag{2.8}$$

From Eq. (2.8), it is inferred that the smallest misalignment between an allowed k_x and a K point is inversely proportional to the diameter. Thus, from the slope of a cone near K points in Eq. (2.4a), the bandgap E_g can be expressed as

$$E_g = 2 \times \left(\frac{\partial E}{\partial k}\right) \times \frac{2}{3d} = 2\hbar v_F \left(\frac{2}{3d}\right) \approx 0.7\text{eV}/d \,(\text{nm}) \tag{2.9}$$

Therefore, semiconducting CNTs ($d = 0.8$–3 nm) exhibit bandgap ranging from 0.9 to 0.2 eV. Depending on the value of $\bar{p}$, where $\bar{p}$ is the remainder when p and q is divided by 3, SWCNTs [represented by (p, q)] can be of three types:

$\bar{p} = 0$; metallic with linear subbands crossing at the K points.
$\bar{p} = 1, 2$; semiconducting with a bandgap, $E_g \sim 0.7$ eV/d (nm).

Similar treatment can also be applied for armchair CNTs $(\bar{p}, \bar{p})$, arriving at the conclusion that they are always metallic.

2.2.3 Properties and Characteristics of CNTs

The atomic arrangements of carbon atoms are responsible for the unique electrical, thermal, and mechanical properties of CNTs [35, 36]. The sp^2 bonding delivers the high conductivity and mechanical strengths to the CNTs. The unique properties of CNTs are discussed below.

2.2.3.1 Strength and Elasticity

Due to the sp^2-hybridization, each carbon atom in a single sheet of graphite is connected via strong sigma bonds to three neighboring atoms. Thus, CNTs exhibit the strongest basal plane elastic modulus and hence are expected to be the ultimate high-strength fiber. The elastic modulus of CNT is much higher than steel that makes the CNT as a strongest material. Although forcing on the tip of nanotube will cause it to bend, the nanotube returns to its original state as soon as the force is removed. This property makes CNTs extremely useful as probe tips for high-resolution scanning probe microscopy. Although, the current Young's modulus of SWCNT is about 1 TPa, but a much higher value of 1.8 TPa has also been reported [37]. For different experimental measurement techniques, the values of Young's modulus vary in the range of 1.22–1.26 TPa depending on the size and chirality of the SWCNTs [36]. It has been observed that the elastic modulus of CNTs is not strongly dependent on the diameter. Primarily, the moduli of CNTs correlate to the amount of disorder in the nanotube walls [38].

2.2.3.2 Thermal Conductivity and Expansion

CNTs can exhibit superconductivity below 20 K (−253 °C) due to the strong in-plane sigma bonds in between carbon atoms. The sigma bond provides exceptional strength and stiffness against axial strains. Moreover, the larger interplane and zero in-plane thermal expansion of SWCNTs results in high flexibility against nonaxial strains.

Due to their high thermal conductivity and large in-plane expansion, CNTs exhibit exciting prospects in nanoscale molecular electronics, sensing and actuating devices, reinforcing additive fibers in functional composite materials, etc. Recent experimental measurements suggest that the CNT-embedded matrices are stronger in comparison to bare polymer matrices [39]. Therefore, it is expected that the nanotube may also significantly improve the thermomechanical and the thermal properties of the composite materials.

2.2.3.3 Field Emission

Under the application of strong electric field, tunneling of electrons from the metal tip to vacuum results in the phenomenon of field emission. Field emission results from the high aspect ratio and small diameter of CNTs. The field emitters are suitable for the application in flat panel displays. For multiwalled CNTs, the field emission properties occur due to the emission of electrons and light. Without applied potential, the luminescence and light emission occur through the electron field emission and visible part of the spectrum, respectively.

2.2.3.4 Aspect Ratio

One of the exciting properties of CNTs is the high aspect ratio, which infers that a lower CNT load is required compared to other conductive additives to achieve similar electrical conductivity. The high aspect ratio provides unique electrical conductivity in CNTs in comparison to the conventional additive materials such as chopped carbon fiber, carbon black, or stainless steel fiber.

2.2.3.5   Absorbent

Carbon nanotubes and CNT composites have been emerging as perspective absorbing materials due to their light weight, larger flexibility, high mechanical strength, and large surface area. Therefore, CNTs emerge out as ideal candidates for use in gas, air, and water filtration. The absorption frequency range of SWCNT–polyurethane composites broaden from 6.4–8.2 (1.8 GHz) to 7.5–10.1 (2.6 GHz) and to 12.0–15.1 GHz (3.1 GHz) [40]. A lot of research has already been carried out for replacing the activated charcoal with CNTs for certain ultra-high purity applications [41].

2.2.3.6   Conductivity

CNTs are assumed to be the most electrically conductive materials. However, it is quite difficult to control the chirality of the SWCNT shells and therefore statistically only one-third of the CNTs in a bundle is assumed to be conducting and the rest of them are semiconducting. However, because of large diameters, the CNT shells of MWCNTs would be conductive even if they are of semiconductor characteristics. The energy gap between the conduction band edge and the Fermi level of a CNT shell is defined as [30]

$$E_g = \frac{v_0 p_{C-C}}{d} \tag{2.10}$$

where d is the CNT diameter, v_0 is the nearest-neighboring tight-binding parameter, and p_{C-C} is the nearest neighbor C–C bond length, which is ~ 0.142 nm. From Eq. (2.10), it can be observed that the bandgap is inversely proportional to the diameter. Therefore, the semiconducting CNT shells with larger diameter are conductive. The detailed conductivity comparison between MWCNTs and SWCNTs will be discussed in the following sections.

2.2.4   Conductivity Comparison

The performance of interconnect primarily depends on the conductivity of the interconnect filler material. The conductivity comparison among Cu, SWCNT, and MWCNT is analyzed in this section.

2.2.4.1 SWCNT Conductivity

The conductivity of SWCNT [42, 43] can be expressed as

$$\sigma_{SWCNT} = \frac{4G_0 l_0 d F_m}{\sqrt{3}(d+\delta)^2} \frac{l}{(l+l_0 d)} \tag{2.11}$$

where l, d, F_m are the interconnect length, shell diameter, fraction of metallic CNTs in the bundle, respectively, l_0 is 10^3, δ is 0.34 nm, G_0 is the quantum conductance equal to $2e^2/h$, e is the charge of an electron, and h is the Planck's constant.

For $l > l_0 d$, Eq. (2.11) can be expressed as

$$\sigma_{SWCNT} \approx \frac{4G_0 l_0 d F_m}{\sqrt{3}(d+\delta)^2} \tag{2.12}$$

From Eq. (2.12), it can be observed that for longer interconnects, the conductivity of SWCNT is independent of length.

2.2.4.2 MWCNT Conductivity

The conductivity of MWCNT [44] can be expressed as

$$\sigma_{MWCNT} = \frac{G_0 l}{2\delta} \left[\left(1 - \frac{d_{min}^2}{d_{max}^2}\right) \frac{a}{2} + \left(b - \frac{l}{l_0}a\right) \times \left(\left(\frac{1}{d_{max}} - \frac{d_{min}}{d_{max}^2}\right) - \frac{l}{d_{max}^2 l_0} \ln \frac{d_{max} + \frac{l}{l_0}}{d_{min} + \frac{l}{l_0}} \right) \right] \tag{2.13}$$

where d_{max} and d_{min} are the outermost and the innermost shell diameters of an MWCNT, respectively; a and b are constants and the values are 0.0612 nm^{-1} and 0.425, respectively [45]. From Eq. (2.13), it can be observed that for $l > (l_0 b/a)$, the conductivity increases with an increase in d_{max}.

The conductivity comparison plot among Cu, SWCNT bundles, and MWCNT is shown in Fig. 2.8. It can be observed that for shorter interconnect length, the conductivity of SWCNT bundle is higher than MWCNT, whereas for longer lengths, MWCNTs can potentially have conductivities several times larger than that of copper or even SWCNT, which is essential for interconnect applications. It is worth noting that the best case scenario is considered for SWCNTs, wherein they were densely packed so that highest conductivity is obtained. However, in contrast to this, an average case scenario was considered for MWCNTs wherein the innermost diameter is half of the outermost shell diameter. The innermost diameters were considered as 5, 15, 35, 50 nm for respective outer diameters of 10, 30, 70, 100 nm. However, the best case scenario for MWCNTs would have been when the innermost diameter had been 1 nm. But still, for longer interconnects, the performance of the MWCNTs was better than the SWCNTs.

Fig. 2.8 The conductivity comparison among Cu, SWCNT, and MWCNT

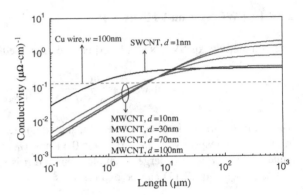

2.2.5   MWCNT Interconnect Modeling

MWCNTs have recently acquired importance for VLSI on-chip interconnect material due to their high current carrying capabilities. For the first time, Burke [46] proposed an electrical equivalent model for the analysis of CNT interconnects based on the Lüttinger liquid theory. The model considered the quantum effects of a nanowire by including the quantum resistance, kinetic inductance, and quantum capacitance. The electrical equivalent model was further explained by Avouris et al. [47] through extensive study of electronic structure and transport properties of CNTs. Depending on the analysis, a bottom-up approach was demonstrated by Li et al. [48] to integrate MWCNTs into multilevel interconnects in silicon-integrated circuits. Ngo et al. [49] reported the mechanism of electron transport across metal–CNT interface. The authors analyzed this mechanism for two different MWCNT architectures, horizontal or side-contacted MWCNTs and vertical or end-contacted MWCNTs. Later, Miano and Villone [50] extended the fluid theory model for frequency domain to describe the electromagnetic response of three-dimensional (3D) structures formed by metallic CNTs and conductors within the framework of classical electrodynamics.

Xu and Srivastava [51] presented a semiclassical one-dimensional (1D) electron fluid model that took into account the electron–electron repulsive force. Based on the one-dimensional electron fluid theory, the authors presented a transmission line model of metallic CNT interconnects using classical electrodynamics. However, the authors neglected the inter-CNT tunneling phenomenon. Later, Li et al. [43] presented a multiconductor transmission line (MTL) model for the MWCNT. The authors considered the tunneling effect between the adjacent shells in MWCNT and neighboring CNTs in a bundle. However, using the MTL model, the analysis of MWCNT with N number of tubes leads to the solution of differential equations with the system dimensional of $2N$, which can be computationally expensive. For this reason, the equivalent single conductor (ESC) model was proposed in [52].

The ESC model is based on the assumption that voltages at an arbitrary cross section along MWCNT are the same, such that all nanotubes are connected in parallel at both ends. The accuracy of the ESC model in comparison to MTL model has been reported by several researchers [52, 53]. It was observed that the transient responses to a pulse input of MTL model and ESC model are in good agreement. The MTL and ESC models are briefly described in the next section.

2.2.5.1   MTL and ESC Models of MWCNT Interconnect

The electrical equivalent models of MWCNT interconnect are discussed in this section. The schematic cross-sectional view of MWCNT interconnect is shown in Fig. 2.9. The MWCNT bundle interconnect line is positioned over a ground plane at a distance H and placed in a dielectric medium with dielectric constant ε. The MWCNT interconnect consists of N number of tubes with intershell distance δ, inner shell diameter d_1, and outer shell diameter d_N. The total number of CNTs in an MWCNT can be expressed as

$$N = 1 + \text{int}\left[\frac{(d_N - d_1)}{2\delta}\right] \qquad (2.14)$$

where int[·] represents an integer value. The number of conducting channels in a CNT can be derived by adding all the subbands contributing to the current conduction. Using Fermi function, it can be expressed as

$$N_{\text{ch},i} = \sum_{\text{subbands}} \frac{1}{\exp(|E_i - E_F|/k_B T) + 1} \qquad (2.15)$$

where T is the temperature, k_B is the Boltzmann constant, and E_i is the lowest (or highest) energy for the subbands above (or below) the Fermi level E_F.

The multiconductor transmission line (MTL) model of MWCNT interconnect is described in Fig. 2.10, where $R_{MC,i}$ and $R_{Q,i}$ represent the imperfect metal contact

Fig. 2.9 Geometry of an MWCNT with N shells above a ground plane

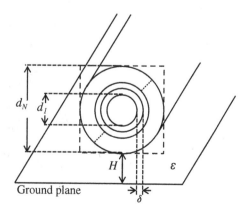

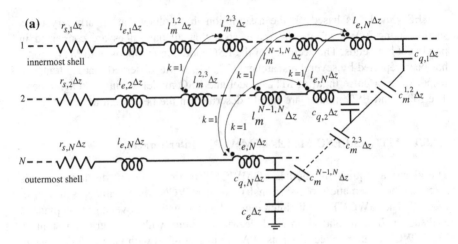

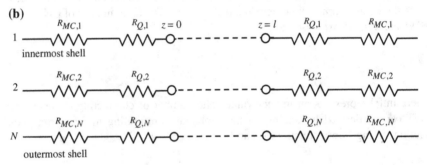

Fig. 2.10 Multiconductor transmission line model of MWCNT (**a**) section of infinitesimal length Δz, where $k = 1$ represents perfect magnetic coupling (**b**) nanotube of length l including terminal resistance

resistance and quantum resistance of ith shell, respectively; $r_{s,i}$, $l_{k,i}$, and $c_{q,i}$ represent the p.u.l. scattering resistance, kinetic inductance, and quantum capacitance, respectively. The parasitics $R_{Q,i}$, $r_{s,i}$, $l_{k,i}$, and $c_{q,i}$ can be expressed as

$$R_{Q,i} = \frac{h}{4e^2 N_{ch,i}} \qquad (2.16a)$$

$$r_{s,i} = \frac{h}{2e^2 \lambda_{mfp,i} N_{ch,i}} \qquad (2.16b)$$

$$l_{k,i} = \frac{h}{4e^2 v_F N_{ch,i}} \qquad (2.16c)$$

$$c_{q,i} = \frac{4e^2 N_{ch,i}}{h v_F} \qquad (2.16d)$$

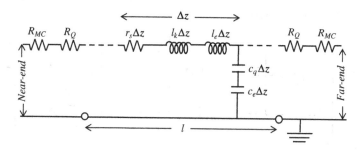

Fig. 2.11 Equivalent single conductor model of MWCNT

where h, e λ_{mfp}, and v_F represent the Planck's constant, electron charge, mean free path, and Fermi velocity, respectively.

In Fig. 2.10, $l_{e,i}$ is p.u.l. magnetic inductance of the ith shell and c_e is p.u.l. electrostatic capacitance, $c_m^{i,i+1}$ and $l_m^{i,i+1}$ are the p.u.l. coupling capacitance and mutual inductance between the shells, respectively. These parasitics can be expressed as

$$l_{e,i} = \frac{\mu_0 \mu_r}{2\pi} \cosh^{-1}\left[\left(\frac{d_i + 2H}{d_i}\right)\right] \tag{2.16e}$$

$$c_e = \frac{2\pi\varepsilon_0\varepsilon_r}{\cosh^{-1}\left[\left(\frac{d_N + 2H}{d_N}\right)\right]} \tag{2.16f}$$

$$c_m^{i,i+1} = \frac{2\pi\varepsilon_0}{\ln(d_{i+1}/d_i)}, \quad i = 1, 2, \ldots, N-1 \tag{2.16g}$$

$$l_m^{i,i+1} = \frac{\mu}{2\pi}\ln(d_{i+1}/d_i), \quad i = 1, 2, \ldots, N-1 \tag{2.16h}$$

To reduce the complexity of the MTL model, a simplified ESC model was proposed in [52]. The ESC model is shown in Fig. 2.11. This model was developed based on the assumption that voltages at an arbitrary cross section along MWCNT are the same. Thus, all the scattering resistances $r_{s,i}$ are in parallel and can be replaced by an equivalent resistance ($r_{s,\mathrm{ESC}}$). The $r_{s,\mathrm{ESC}}$ can be expressed as

$$r_{s,\mathrm{ESC}} = \frac{h/e^2}{\sum_{i=1}^{N} 2N_{\mathrm{ch},i}\lambda_{\mathrm{mfp},i}} \tag{2.17a}$$

Referring to Fig. 2.12, the distributed MWCNT capacitance $c_{q,\mathrm{ESC}}$ is expressed in terms of quantum capacitance and coupling capacitance between shell to shell.

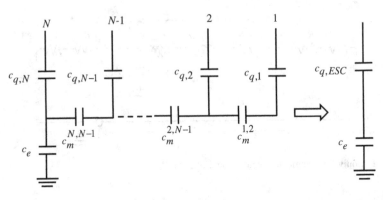

Fig. 2.12 Per unit length capacitance network of the MWCNT

$$c_{\text{equ},1} = c_{q,1} \tag{2.17b}$$

$$c_{\text{equ},i} = \left(\frac{1}{c_{\text{equ},i-1}} + \frac{1}{c_m^{i-1,i}} \right)^{-1} + c_{q,i}, \quad i = 2, 3, \dots, N \tag{2.17c}$$

$$c_{q,\text{ESC}} = c_{\text{equ},N} \tag{2.17d}$$

The inductance equations can be written in a similar form. The ESC model of MWCNT interconnect is thoroughly discussed in Sect. 4.2.

2.2.6 MWCNT Performance Analysis

The performance analysis of MWCNT interconnects was analyzed using both the MTL and ESC models in [54]. The voltage response of two coupled MWCNT interconnects of 14 and 22 nm technologies was computed to a pulse input. It was observed that both models are in good agreement. The same agreement was achieved in the estimation of 50 % time delay as well. The validity of the ESC model was also verified experimentally in [52]. Based on the ESC model, Lamberti et al. [55] compared the performance of MWCNTs with SWCNTs. The propagation delay time was analyzed at three different technologies 15, 21, and 32 nm by means of interval analysis. It was observed that for global interconnect lengths, the time delay obtained for MWCNT interconnects is less than 1 ns for the most severe configuration, i.e., for 15 nm technology node at a length of 250 μm, whereas for SWCNTs the delay is as large as 7.87 ns.

The estimation of performance parameters under the crosstalk influence is an important design concern in modern VLSI interconnects. The crosstalk analysis of MWCNTs has been studied by several researchers. Liang et al. [56] analyzed the crosstalk noise effects with lengths ranging from 10 to 1000 μm at 22 and 14 nm technology nodes. Moreover, the performance of MWCNTs was compared with the Cu interconnects. They observed that the MWCNT interconnects showed better performance for longer wire lengths and smaller technology nodes. Das et al. [57] analyzed the crosstalk effects in Cu, SWCNT, and MWCNT interconnects. They observed that the MWCNT-based interconnects are more suitable for VLSI interconnects. Furthermore, the authors analyzed the power supply voltage drop for Cu- and MWCNT-based interconnects in [58]. It was observed that the CNT-based interconnects have significantly less power drop in comparison to that of Cu-based interconnects for semi-global and global lengths. Based on the ESC model, Liang et al. [59] investigated the crosstalk effects in Cu and MWCNT interconnects. They reported that the crosstalk-induced time delays in MWCNT interconnects are much smaller than those in the Cu interconnects. Sahoo and Rahaman [60] developed an analytical closed-form delay expression for both Cu and MWCNT interconnects. They observed that the performance of MWCNT interconnects over copper interconnects is improved by 90 % for 200 µm long interconnect. In 2015, Tang et al. [61] proposed a fast transient simulation technique based on the ESC model for the crosstalk-induced performance analysis of MWCNT interconnects. They observed that the proposed method and HSPICE are very similar to each other with an average relative error of 1.54 %. However, most of the researchers [56, 59–61] used the resistive driver in the performance analysis of MWCNT interconnects that leads to severe errors in the performance estimation of the driver-interconnect-load (DIL) systems.

2.3   Graphene Nanoribbons

In 1996, Mitsutaka Fujita and his group provided a theoretical model of graphene nanoribbons (GNRs) to observe the edge and nanoscale dimension effect in graphene [62, 63]. Recent developments in GNRs have aroused a lot of research interest of their potential applications in the area of interconnects and field-effect transistors [64–66]. A monolithic system can be constructed using the single-layer GNR for both transistors and interconnects. For nanoscale device dimensions, Cu-based interconnects are mostly affected by grain boundaries and sidewall scatterings. It has been predicted that GNRs will outperform the Cu interconnects for smaller widths [67]. In a high-quality GNR sheet, the mean free path is ranging from 1 to 5 µm. GNRs can carry large current densities of more than 10^8 A/cm^2. They also offer high carrier mobility that can reach up to 10^5 cm^2/(Vs) [68].

2.3.1   Basic Structure of GNRs

A graphene nanoribbon is a single sheet of graphene layer, which is extremely thin and limited in width, such that it results in a one-dimensional structure [69]. As a

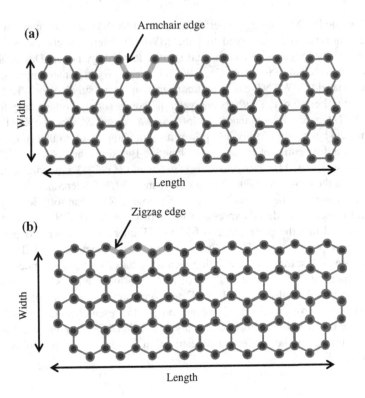

Fig. 2.13 Structure of GNR **a** armchair and **b** zigzag

result, GNRs can be considered as an unrolled version of CNTs. The electronic properties of GNRs are similar to that of CNTs. Depending on termination of their width, GNRs can be divided into chiral and nonchiral GNRs. The chiral GNRs can be further classified as armchair (ac) or zigzag (zz) GNRs as shown in Fig. 2.13a, b, respectively. It can be noted that the terms "armchair" and "zigzag" are used for both GNRs and CNTs. However, these nomenclatures are used in opposite ways. For GNRs the terms armchair and zigzag indicate the pattern of the GNR edge, whereas for CNTs the same terms indicate the CNT circumference. Therefore, the unrolled armchair CNT can be visualized as a zigzag GNR and vice versa.

Depending on the stacked graphene sheets, GNRs are classified as single-layer GNR (SLGNR) or multilayer GNR (MLGNR). The most promising interconnect solution for VLSI interconnect is MLGNR due to its higher current carrying capability than SLGNR. The geometric structure of an MLGNR is shown in Fig. 2.14. The MLGNR interconnect consists of N number of layers, with interlayer distance δ, width w, and thickness t.

From the fabrication point of view, it is evident that the growth of the GNRs can be more easily controlled than that of the CNTs because of their planar structure. This makes them compatible with the conventional lithography techniques [70].

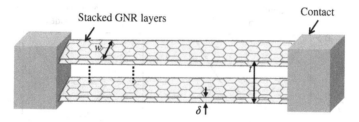

Fig. 2.14 The geometric structure of an MLGNR interconnect

Using the electron beam lithography technique, Murali et al. [71] fabricated an MLGNR interconnect with ten layers. The higher electrical conductivity in MLGNR can be obtained either by enhancing the carrier mobility or by increasing the number of carriers. The carrier mobility can be increased by intercalation doping of arsenic pentafluoride (AsF$_5$) vapor. Using the AsF$_5$ doping, the conductivity of MLGNR can be increased up to 3.2×10^5 S/cm, which is almost 1.5 times higher than the copper interconnects [72]. Additionally, the easier fabrication process of MLGNRs makes them as promising candidates for interconnect material.

2.3.2   Semiconducting and Metallic GNRs

GNRs can act as either semiconducting or metallic based on the pattern of the GNR edge. The zigzag edge-patterned GNR always act as metallic, whereas the armchair edge-patterned GNRs can act as either metallic or semiconducting depending on the number of carbon atoms present across the width of the GNR. This section presents the behavior of armchair GNRs and its dual nature.

The typical structure of armchair GNR is shown in Fig. 2.13a, where the number of carbon atoms across its width, $N_C = 7$. For understanding the metallic/semiconducting behavior of GNRs, it is necessary to analyze the electronic band structures. The band structures of GNRs are obtained using a tight-binding (TB) model [73]. Using the TB approach, the band structures of 23- and 24-atom wide armchair GNRs are shown in Fig. 2.15a, b, respectively. It can be observed that the metallic GNR has zero bandgap, whereas the semiconducting GNR has 0.2 eV bandgap. The ac GNR acts as metallic, if $N_C = 3a + 2$ and acts as semiconducting, if $N_C = 3a + 1$ or 3, where a is an integer. The zz GNRs are always metallic, independent of the value of N_C [73].

2.3.3   Properties and Characteristics of GNRs

Most of the physical and electrical properties of GNRs are similar to that of CNTs. However, compared to CNTs, the growth of the GNRs is considered to be more

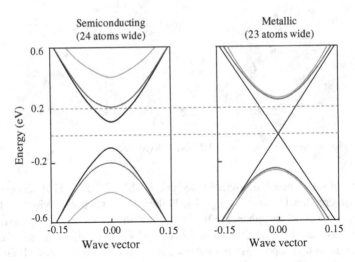

Fig. 2.15 Band structures of **a** semiconducting and **b** metallic armchair GNRs whose widths are 6.02 nm (24 atoms wide) and 5.78 nm (23 atoms wide), respectively

controllable due to their planar structure. Moreover, the major advantage of GNRs over CNTs is that both transistor and interconnect can be fabricated on the same continuous graphene layer, which unlike CNTs, are free from Stone–Wales defects [74]. Therefore, one of the manufacturing difficulties regarding the formation of metal–nanotube contact can be avoided. Due to the lower resistivity, the MLGNRs are often preferred over SLGNRs as suitable on-chip interconnect material. However, the MLGNRs fabricated till date, have displayed some level of edge roughness [70, 71]. The electron scattering at the rough edges reduces the mean free path that substantially lowers the conductance of the MLGNR. This fundamental challenge limits the performance of MLGNR-based interconnects. The value of MFP primarily depends on the level of edge roughness. The following sections discuss the effect of edge roughness on the MFP.

2.3.3.1 Mean Free Path of GNR

The effective MFP of GNR λ_{eff} depends on the scattering effects due to phonons λ_{ph}, defects λ_d, and edge roughness λ_n. Using Matthiessen's rule, the λ_{eff} can be expressed as

$$\frac{1}{\lambda_{\text{eff}}} = \frac{1}{\lambda_d} + \frac{1}{\lambda_n} + \frac{1}{\lambda_{\text{ph}}} \tag{2.18}$$

For the interconnect applications (low bias), the MFP corresponding to λ_{ph} is observed as extremely large, i.e., tens of micrometers, and therefore, its effect can be neglected for the modeling of GNR scattering resistance [73]. Consequently, λ_d and λ_n dominate the overall value of λ_{eff}.

According to the experimental measurements reported in [75], the MFP corresponding to λ_d is about 1 μm for a single-layer GNR, which is width independent. However, in multilayer GNR, the MFP reduces due to the intersheet electron hopping [76]. The λ_d of MLGNR can be extracted by measuring the in-plane conductivity of GNR. Using the in-plane conductivity of $G_{\text{sheet}} = 0.026$ (μΩ-cm)$^{-1}$ [77], layer spacing of 0.34 nm, and $E_F = 0$ of a neutral MLGNR, the λ_d is extracted as 419 nm by solving Eq. (2.19) [77]

$$G_{\text{sheet}} = \frac{2q^2}{h} \cdot \frac{\pi \lambda_d}{h v_f} \cdot 2k_B T \ln\left[2\cosh\left(\frac{E_F}{2k_B T}\right)\right] \tag{2.19}$$

To increase the conductivity of MLGNR, AsF$_5$ intercalated graphite can be used. The in-plane conductivity, $G_{\text{sheet}} = 0.63$ (μΩ-cm)$^{-1}$ and carrier concentration, $n_p = 4.6 \times 10^{20}$ cm^{-3} are observed for the AsF$_5$ intercalated graphite [78]. Using the simplified TB model, the E_F can be expressed as

$$E_F = h v_F \left(\frac{n_p \cdot \delta}{4\pi}\right)^{1/2} \tag{2.20}$$

where $\delta = 0.575$ nm is the average layer spacing between adjacent graphene layers. Using the expressions (2.19) and (2.20), E_F and λ_d are expressed as 0.6 eV and 1.03 μm, respectively.

The MFP corresponds to diffusive scattering at the edges is a function of edge backscattering probability, P and the average distance by an electron travels along the length before hitting the edge. The mean free path for nth subband due to edge scattering can be expressed as [67]

$$\lambda_n = \frac{w}{P} \sqrt{\left(\frac{E_F/\Delta E}{n}\right)^2 - 1} \tag{2.21}$$

where ΔE is the gap between the subbands.

2.3.4 Conductivity Comparison

The performance of the interconnect line is primarily depends on the conductivity of the material. This section discusses the conductivity of various interconnect materials. Figure 2.16 shows the conductivity of Cu, SWCNT bundle, MWCNT and MLGNR interconnects. The fully specular edge MLGNR interconnects are analyzed for two different doped conditions [79]. First, the Fermi energy level of 0.3 eV is considered and second, the level of 0.6 eV is considered. The SWCNT

Fig. 2.16 Conductivity comparison among Cu, SWCNT bundle, MWCNT and MLGNR interconnects

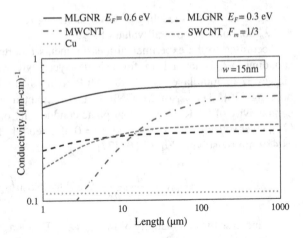

diameter is chosen to be 1 nm with a metallic to semiconducting ratio (F_m) of 1/3. For MWCNTs, the outer diameter of CNT shell is considered as 15 nm. From Fig. 2.16, it can be observed that the conductivity of MLGNR increases with the Fermi energy. Moreover, at highly doped condition ($E_F = 0.6$ eV) the conductivity of MLGNR is observed to be higher than MWCNT interconnects.

2.3.5 MLGNR Interconnect Modeling

This section presents an electrical equivalent model of the MLGNR interconnect line. An MLGNR of width w and thickness t is placed above the ground plane at a distance H as shown in Fig. 2.17. The permittivity of the medium between the bottommost layer of MLGNR and the ground plane is represented by ε. The total number of layers (N_{layer}) can be expressed as

$$N_{\text{layer}} = 1 + \text{int}\left[\frac{t}{\delta}\right] \tag{2.22}$$

Fig. 2.17 Geometry of MLGNR above ground plane

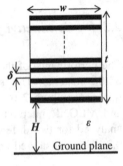

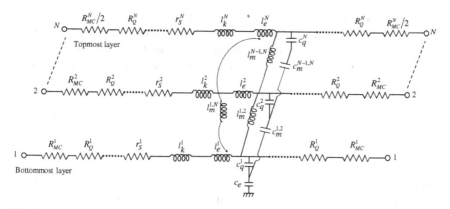

Fig. 2.18 Equivalent *RLC* model of MLGNR interconnect

The interlayer distance δ is considered to be 0.575 and 0.34 nm for doped and neutral MLGNRs [77], respectively.

The equivalent electrical model of MLGNR interconnect is presented in Fig. 2.18, wherein the parasitics are primarily dependent on the number of conducting channels (N_{ch}) of each layer in MLGNR. The N_{ch} takes into account the effect of spin and sublattice degeneracy of carbon atoms and primarily depends on the width, Fermi energy (E_F), temperature (T) and can be expressed as [80]

$$N_{ch} = \sum_{n=0}^{n_C} \left[e^{(E_i - E_F)/kT} + 1 \right]^{-1} + \sum_{n=0}^{n_V} \left[e^{(E_i + E_F)/kT} + 1 \right]^{-1} \qquad (2.23)$$

where k, n_C, and n_V represent the Boltzmann constant, number of conduction and valence bands, respectively. E_i is the lowest/highest energy of ith subband in conduction/valence band [80].

Depending on the current fabrication process, the imperfect metal–MLGNR contact resistance (R_{MC}) has a typical value ranging from 1 to 20 kΩ [81]. Each layer of MLGNR exhibits lumped quantum resistance (R_Q) that is due to the quantum confinement of carriers across the interconnect width. The quantum resistance of jth layer (R_Q^j) can be expressed as

$$R_Q^j = \frac{h}{4e^2 \cdot N_{ch}} \qquad (2.24)$$

For longer interconnects, scattering resistance r_s appears due to the static impurity scattering, defects, line edge roughness scattering (LER), etc. [82–84]. The r_s primarily depends on the effective MFP of electrons (λ_{eff}) and can be expressed as

$$r_s^j = \frac{h}{2e^2 \cdot N_{ch} \cdot \lambda_{eff}} \qquad (2.25)$$

Using the Matthiessen's rule, the λ_{eff} of each subband can be expressed from (2.18). Each layer in MLGNR comprises of kinetic inductance (l_k) and quantum capacitance (c_q) that represent the mobile charge carrier inertia and the density of electronic states, respectively. The l_k and c_q of any layer j can be expressed as

$$l_k^j = \frac{l_{k0}}{2N_{ch}}; \text{ where } l_{k0} = \frac{h}{2e^2 v_F} \tag{2.26}$$

$$c_q^j = 2c_{q0} \cdot N_{ch}; \text{ where } c_{q0} = \frac{2e^2}{hv_F} \tag{2.27}$$

where $v_F \approx 8 \times 10^5$ m/s represents the Fermi velocity of carriers in graphene [81]. The kinetic inductance per channel is 8 nH/μm, which is verified by the experimental observations also [85]. The electrostatic capacitance (c_e) is due to the electric field coupling between the bottom most layer of MLGNR and the ground plane. Therefore, the c_e is primarily dependent on the MLGNR width (w) and the distance (H) from the ground plane. Apart from this, the magnetic inductance (l_e) of MLGNR interconnect is due to the stored energies of carriers in the magnetic field. The l_e and c_e can be expressed as

$$l_e^j = \frac{\mu_0 \mu_r H}{w} \quad \text{and} \quad c_e = \frac{\varepsilon_0 \varepsilon_r w}{H} \tag{2.28}$$

The interlayer mutual inductance (l_m) and coupling capacitance (c_m) are mainly due to the magnetic and electric field coupling between the adjacent layers. The l_m and c_m can be expressed as

$$l_m^{j-1,j} = \frac{\mu_0 \delta}{w}, \quad j = 2, 3, \ldots, N \tag{2.29a}$$

$$c_m^{j-1,j} = \frac{\varepsilon_0 w}{\delta}, \quad j = 2, 3, \ldots, N \tag{2.29b}$$

The analysis of signal propagation along an MLGNR with N_{layer} leads to the solution of a 2N-dimensional system of differential equations that can be highly time-consuming. For this reason, the equivalent RLC model of Fig. 2.18 is simplified to an ESC model shown in Fig. 2.19, wherein all the layers are assumed to be parallel. The value of $R_1 = (R_{MC} + R_Q)$ is equally divided between the two contacts on either side of the interconnect line. The detailed explanation of ESC model of MLGNR interconnect line is provided in Sect. 5.2.

2.3.6 MLGNR Performance Analysis

The performance of an MLGNR interconnect is generally evaluated by means of an electrical equivalent model. The equivalent model considers all the parasitic

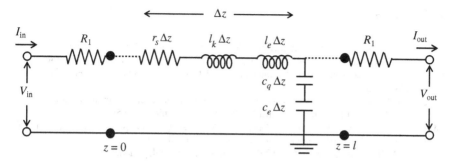

Fig. 2.19 Equivalent single conductor (ESC) model of MLGNR interconnect

parameters based on the quantum effects of the nanowire, and its electrostatic and magnetostatic characteristics. Sarto et al. proposed an electrical equivalent transmission model to represent the MLGNR interconnect [86]. They compared the performance between MWCNT and MLGNR interconnects and observed that the MLGNR interconnect has higher current carrying capability than the MWCNT interconnect. Xu et al. [77] derived the conductance model of MLGNR interconnect using the tight-binding approach and the Landauer formula. The conductance of the MLGNR is compared among Cu, W, and CNTs. They observed that the conductance of MLGNR is much higher than Cu, W, and CNTs when proper intercalation-doped MLGNRs are used. Nishad et al. [87] presented the analytical time domain models for the performance analysis of top contact and side contact MLGNR interconnects. Based on the analytical models, they designed an optimum top-contacted MLGNR interconnect that exceeds the performance of Cu and optical interconnects.

The crosstalk-induced signal transmission analysis of MLGNR interconnects was performed by Cui et al. [81] based on the transmission line model. The authors obtained the output response of driver-interconnect-load system using the transfer function. The impact of Fermi level on the signal transmission was also investigated. In 2014, Zhao et al. [79] performed the comparative study on MLGNR interconnects with SWCNT, MWCNT, and Cu interconnects. They observed that even with the maximum crosstalk impacts considered, the advantage of MLGNR interconnects over other interconnect materials can still be maintained. The impact of MLGNR line resistance variations on the crosstalk-induced performance parameters were investigated in [88]. The simulations were performed for 11 and 8 nm technology nodes for both intermediate and global interconnect lengths. They observed that irrespective of technology node, the perfectly doped fully specular MLGNR interconnects are better than Cu interconnects as far as the line resistance tolerance was concerned. However, the existing crosstalk noise models [79, 81, 88] analyzed the performance of MLGNR interconnects with resistive drivers that limits the accuracy of the models. Moreover, the authors considered the mean free path parameter independent of width by assuming perfectly smooth edges of MLGNRs.

References

1. Livshits P, Sofer S (2012) Aggravated electromigration of copper interconnection lines in ULSI devices due to crosstalk noise. IEEE Trans Device Mater Reliab 12(2):341–346
2. Moll F, Roca M, Rubio A (1998) Inductance in VLSI interconnection modeling. IEEE Proc Circuits Devices Syst 145(3):175–179
3. Sakurai T, Newton R (1990) Alpha-power law MOSFET model and its applications to CMOS inverter delay and other formulas. IEEE J Solid-State Circuits 25(2):584–594
4. Bowman KA, Austin BL, Eble JC, Xingha T, Meindl JD (1999) A physical alpha-power law MOSFET model. IEEE J Solid State Circuits 34(10):1410–1414
5. Dutta S, Shetti SSM, Lusky SL (1995) A comprehensive delay model for CMOS inverters. IEEE J Solid-State Circuits 30(8):864–871
6. Bisdounis L, Nikolaidis S, Koufopavlou O (1998) Analytical transient response and propagation delay evaluation of the CMOS inverter for short-channel devices. IEEE J Solid-State Circuits 33(2):302–306
7. Qian J, Pullela S, Pillage L (1994) Modeling the effective capacitance for the RC interconnect of CMOS gates. IEEE Trans Comput Aided Des 13(12):1526–1535
8. Hafed M, Oulmane M, Rumin NC (2001) Delay and current estimation in a CMOS inverter with an RC load. IEEE Trans Comput Aided Des 20(1):80–89
9. Chatzigeorgiou A, Nikolaidis S, Tsoukalas I (2001) Modeling CMOS gates driving RC interconnect loads. IEEE Trans Circuits Syst II Analog Digital Signal Process 48(4):413–418
10. Adler V, Friedman EG (1998) Repeater design to reduce delay and power in resistive interconnect. IEEE Tran Circuits Syst II Analog Digital Signal Process 45(5):607–616
11. Bakoglu HB, Meindl JD (1985) Optimal interconnection circuits for VLSI. IEEE Trans Electron Devices 32(5):903–909
12. Rubinstein J, Penfield P, Horowitz MA (1983) Signal delay in RC tree networks. IEEE Trans Comput Aided Des 2(3):202–211
13. Kahng A, Muddu S (1997) An analytical delay model for RLC interconnects. IEEE Trans Comput Aided Des 16(2):1507–1514
14. Ismail YI, Friedman EG, Neves JL (2000) Equivalent elmore delay for RLC trees. IEEE Trans Comput Aided Des 19(1):83–97
15. Bai X, Chandra R, Dey S, Srinivas PV (2004) Interconnect couplingaware driver modeling in static noise analysis for nanometer circuits. IEEE Trans Comput Aided Des 23(8):1256–1263
16. Davis JA, Meindl JD (2000) Compact distributed RLC models, part I: single line transient, time delay, and overshoot expressions. IEEE Trans Electron Devices 47(11):2068–2077
17. Davis JA, Meindl JD (2000) Compact distributed RLC models, part II: coupled line transient expressions and peak crosstalk in multilevel networks. IEEE Trans Electron Devices 47(11):2078–2087
18. Agarwal K, Sylvester D, Blaauw D (2006) Modeling and analysis of crosstalk noise in coupled RLC interconnects. IEEE Trans Comput Aided Des Integr Circuits Syst 25(5):892–901
19. Liu T, Kuo J, Zhang S (2012) A closed-form analytical transient response model for on-chip distortion less interconnect. IEEE Trans Electron Devices 59(12):3186–3192
20. Kaushik BK, Sarkar S (2008) Crosstalk analysis for a CMOS gate driven inductively and capacitively coupled interconnects. Microelectron J 39(12):1834–1842
21. Kaushik BK, Sarkar S, Agarwal RP, Joshi RC (2010) An analytical approach to dynamic crosstalk in coupled interconnects. Microelectron J 41(2):85–92
22. Li XC, Ma JF, Swaminathan M (2011) Transient analysis of CMOS gate driven RLGC interconnects based on FDTD. IEEE Trans Comput Aided Des Integr Circuits Syst 30(4):574–583
23. Kroto HW, Heath JR, O'Brien SC, Curl RF, Smalley RE (1985) C_{60}: buckminsterfullerene. Nature 318:162–163
24. Scarselli M, Castrucci P, Crescenzi M (2012) Electronic and optoelectronic nano-devices based on carbon nanotubes. J Phys Condens Matter 24(31):313202-1–313202-36

25. Xu T, Wang Z, Miao J, Chen X, Tan CM (2007) Aligned carbon nanotubes for through-wafer interconnects. Appl Phys Letts 91(4):042108-1–042108-3

26. Monthioux M, Serp P, Flahaut E (2010) Introduction to carbon nanotubes. In: Bhushan B (ed) Handbook of nano-technology. Springer, New York

27. Wang N, Tang ZK, Li GD, Chen JS (2000) Single-walled 4 Å carbon nanotube arrays. Nature 408:50–51

28. Javey A, Kong J (2009) Carbon nanotube electronics. Springer

29. Hamada N, Sawada SI, Oshiyama A (1992) New one-dimensional conductors, graphite microtubules. Phys Rev Lett 68:1579–1581

30. Li HJ, Lu WG, Li JJ, Bai XD, Gu CZ (2005) Multichannel ballistic transport in multiwall carbon nanotubes. Phys Rev Lett 95(8):86601

31. Nihei M, Kondo D, Kawabata A (2005) Low-resistance multi-walled carbon nanotube vias with parallel channel conduction of inner shells. In: Proceedings of the IEEE international interconnect technology conference, pp 234–36

32. Forró L, Schönenberger C (2000) Physical properties of multi-wall nanotubes in topics in applied physics, carbon nanotubes: synthesis, structure, properties and applications. In: Dresselhaus MS, Dresselhaus G, Avouris P (eds) Springer-Verlag, Berlin, Germany

33. Wei BQ, Vajtai R, Ajayan PM (2001) Reliability and current carrying capacity of carbon nanotubes. Appl Phys Lett 79(8):1172–1174

34. Close GF, Wong HSP (2008) Assembly and electrical characterization of multiwall carbon nanotube interconnects. IEEE Trans Nanotechnol 7(5):596–600

35. Shah TK, Pietras BW, Adcock DJ, Malecki HC, Alberding MR (2013) Composites comprising carbon nanotubes on fiber. US Patent, US8585934 B2

36. Dresselhaus M, Dresselhaus G, Avouris Ph (2001) Carbon nanotubes: synthesis, structure, properties and applications. Top Appl Res 80

37. Hsieh JYL, Huang JM, Hwang CC (2006) Theoretical variations in the young's modulus of single-walled carbon nanotubes with tube radius and temperature: a molecular dynamics study. Nanotechnology 17:3920–3924

38. Forro L, Salvetat JP, Bonard J (2002) Electronic and mechanical properties of carbon nanotubes. In: Tománek D, Enbody RJ (eds) Science and application of nanotubes. Plenum Publishers, New York, pp 297–320

39. Wei C, Srivastava D, Cho K (2002) Thermal expansion and diffusion coefficients of carbon nanotube-polymer composites. Nano Lett 2(6):647–650

40. Wang Z, Zhao GL (2013) Microwave absorption properties of carbon nanotubes-epoxy composites in a frequency range of 2-20 GHz. Open J Compos Mater 3(2):17–23

41. Ifeanyi HN, John IE, Zhou W, Diola B, Guang-Lin Z (2015) Microwave absorption properties of multi-walled carbon nanotube (outer diameter 20–30 nm)–epoxy composites from 1 to 26.5 GHz. Diam Relat Mater 52:66–71

42. Srivastava A, Xu Y, Sharma AK (2010) Carbon nanotubes for next generation very large scale integration interconnects. J Nanophotonics 4(1):1–26

43. Li H, Yin WY, Banerjee K, Mao JF (2008) Circuit modeling and performance analysis of multi-walled carbon nanotube interconnects. IEEE Trans Electron Devices 55(6):1328–1337

44. Naeemi A, Meindl JD (2006) Compact physical models for multiwall carbon-nanotube interconnects. IEEE Electron Device Lett 27(5):338–340

45. Naeemi A, Sarvari R, Meindl JD (2005) Performance comparison between carbon nanotube and copper interconnects for gigascale integration (GSI). IEEE Electron Device Lett 26(2): 84–86

46. Burke PJ (2002) Lüttinger liquid theory as a model of the gigahertz electrical properties of carbon nanotubes. IEEE Trans Nanotechnol 1(3):129–144

47. Avouris P, Appenzeller J, Martel R, Wind SJ (2003) Carbon nanotube electronics. Proc IEEE 91(11):1772–1784

48. Li J, Ye Q, Cassell A, Ng HT, Stevens R, Han J, Meyyappan M (2003) Bottom-up approach for carbon nanotube interconnects. Appl Phys Lett 82(15):2491–2493

49. Ngo Q, Petranovic D, Krishnan S, Cassell AM, Ye Q, Li J, Meyyappan M, Yang CY (2004) Electron transport through metal–multiwall carbon nanotube interfaces. IEEE Trans Nanotechnol 3(2):311–317

50. Miano G, Villone F (2005) An integral formulation for the electrodynamics of metallic carbon nanotubes based on a fluid model. IEEE Trans Antennas Propag 54(10):2713–2724

51. Xu Y, Srivastava A (2009) A model for carbon nanotube interconnects. Int J Circuit Theory Appl 38(6):559–575

52. Sarto MS, Tamburrano A (2010) Single conductor transmission-line model of multiwall carbon nanotubes. IEEE Trans Nanotechnol 9(1):82–92

53. Tang M, Lu J, Mao J (2012) Study on equivalent single conductor model of multi-walled carbon nanotube interconnects. In: Proceedings of the IEEE Asia Pacific microwave conference, Taiwan, pp 1247–1249

54. D'Amore M, Sarto MS, Tamburrano A (2010) Fast transient analysis of next-generation interconnects based on carbon nanotubes. IEEE Trans Electromagn Compat 52(2):496–503

55. Lamberti P, Tucci V (2012) Impact of variability of the process parameters on CNT-based nanointerconnects performances: a comparison between SWCNTs bundles and MWCNT. IEEE Trans Nanotechnol 11(5):924–933

56. Liang F, Lin H, Wang G (2010) Prediction of crosstalk effects in future multiwall carbon nanotube (MWCNT) interconnects. In: Proceedings of the IEEE symposium on antennas propagation and EM theory (ISAPE), Guangzhou, pp 1031–1034

57. Das D, Rahaman H (2011) Analysis of crosstalk in single- and multiwall carbon nanotube interconnects and its impact on gate oxide reliability. IEEE Trans Nanotechnol 10(6):1362–1370

58. Das D, Rahaman H (2011) IR drop analysis in single- and multi-wall carbon nanotube power interconnects in sub-nanometer designs. In: Proceedings of the IEEE Asia symposium on quality electronic design (ASQED), pp 174–183

59. Liang F, Wang G, Lin H (2012) Modeling of crosstalk effects in multiwall carbon nanotube interconnects. IEEE Trans Electromagn Compat 54(1):133–139

60. Sahoo M, Rahaman H (2013) Performance analysis of multiwalled carbon nanotube bundles. In: Electronics and Nanotechnology (ELNANO), IEEE XXXIII international scientific conference, pp 200–204

61. Tang M, Mao J (2015) Modeling and fast simulation of multiwalled carbon nanotube interconnects. IEEE Trans Electromagn Compat 57(2):232–240

62. Fujita M, Wakabayashi K, Nakada K, Kusakabe K (1996) Peculiar localized state at zigzag graphite edge. J Phys Soc Jpn 65(7):1920–1923

63. Nakada K, Fujita M, Dresselhaus G, Dresselhaus MS (1996) Edge state in graphene ribbons: nanometer size effect and edge shape dependence. Phys Rev 54(24):17954–17961

64. Echtermeyer TJ, Lemme MC, Baus M, Szafranek BN, Geim AK, Kurz H (2008) Nonvolatile switching in graphene field-effect devices. IEEE Electron Device Lett 29(8):952–954

65. Lemme MC, Echtermeyer TJ, Baus M, Kurz H (2007) A graphene field-effect device. IEEE Electron Device Lett 28(4):282–284

66. Rawat B, Paily R (2015) Analysis of graphene tunnel field-effect transistors for analog/RF applications. IEEE Trans Electron Devices 62(8):2663–2669

67. Naeemi A, Meindl JD (2007) Conductance modeling for graphene nanoribbon (GNR) interconnects. IEEE Electron Device Lett 28(5):428–431

68. Li H, Xu C, Srivastava N, Banerjee K (2009) Carbon nanomaterials for next-generation interconnects and passives: physics, status and prospects. IEEE Trans Electron Devices 56(9):1799–1821

69. Kan, E.; Li, Z.; Yang. J. "Graphene nanoribbons: geometric electronic and magnetic properties," In *Physics and Applications of Graphene—Theory*, INTECH, ed. S. Mikhailov, Chapter 16, 2011.

70. Avouris P (2010) Graphene: electronic and photonic properties and devices. Nano Lett 10(11):4285–4294

71. Murali KH, Brenner K, Yang Y, Beck T, Meindl JD (2009) Resistivity of graphene nanoribbon interconnects. IEEE Electron Device Lett 30(6):611–613

72. Dresselhaus MS, Dresselhaus G (2002) Intercalation compounds of graphite. Adv Phys 51 (1):1–186

73. Naeemi A, Meindl JD (2009) Compact physics-based circuit models for graphene nanoribbon interconnects. IEEE Trans Electron Devices 56(9):1822–1833

74. Stan MR, Unluer D, Ghosh A, Tseng F (2009) Graphene devices, interconnect and circuits—challenges and opportunities. In: Proceedings of the IEEE international symposium on circuits and systems (ISCAS), Taipei, pp 69–72

75. Berger C, Song Z, Li X, Wu X, Brown N, Naud C, Mayou D, Li T, Hass J, Marchenkov AN, Conrad EH, First PN, Heer WA (2006) Electronic confinement and coherence in patterned epitaxial graphene. Science 312(5777):1191–1196

76. Benedict LX, Crespi VH, Louie SG, Cohen ML (1995) Static conductivity and superconductivity of carbon nanotubes—Relations between tubes and sheets. Phys Rev B Condens Matter 52(20):14935–14940

77. Xu C, Li H, Banerjee K (2009) Modeling, analysis, and design of graphene nanoribbon interconnects. IEEE Trans Electron Devices 56(8):1567–1578

78. Hanlon LR, Falardeau ER, Fischer JE (1977) Metallic reflectance of AsF_5-graphite intercalation compounds. Solid State Commun 24(5):377–381

79. Wen-Sheng Zhao; Wen-Yan Yin (2014) Comparative study on multilayer graphene nanoribbon (MLGNR) interconnects. IEEE Trans Electromagn Compat 56(3):638–645

80. Nasiri SH, Faez R, Moravvej-Farshi MK (2012) Compact formulae for number of conduction channels in various types of grapheme nanoribbons at various temperatures. Mod Phys Lett B 26(1):1150004-1–115004-5

81. Cui J, Zhao W, Yin W, Hu J (2012) Signal transmission analysis of multilayer graphene nano-ribbon (MLGNR) interconnects. IEEE Trans Electromagn Compat 54(1):126–132

82. Areshkin DA, Gunlycke D, White CT (2007) Ballistic transport in graphene nanostrips in the presence of disorder: importance of edge effects. Nano Lett 7(1):204–210

83. Hwang EH, Adam S, Sarma SD (2007) Carrier transport in two-dimensional graphene layers. Phys Rev Lett 98(18):186806-1–186806-4

84. Yan J, Zhang Y, Kim P, Pinczuk A (2007) Electric field effect tuning of electron-phonon coupling in graphene. Phys Rev Lett 98(16):166802-1–166802-4

85. Plombon JJ (2007) High-frequency electrical properties of individual and bundled carbon nanotubes. Appl Phys Lett 90(6):063106-1–063106-3

86. Sarto MS, Tamburrano A (2010) Comparative analysis of TL models for multilayer graphene nanoribbon and multiwall carbon nanotube interconnects. In: Proceedings of the IEEE international symposium on electromagnetic compatibility, Fort Lauderdale, FL, USA, pp 212–217

87. Nishad AK, Sharma R (2014) Analytical time-domain models for performance optimization of multilayer GNR interconnects. IEEE J Sel Top Quantum Electron 20(1):3700108-1–3700108-8

88. Sahoo M, Rahaman H (2014) Impact of line resistance variations on crosstalk delay and noise in multilayer graphene nano ribbon interconnects. In: Proceedings of the international symposium on electronic system Design (ISED), pp 94–98

Chapter 3
FDTD Model for Crosstalk Analysis of CMOS Gate-Driven Coupled Copper Interconnects

Abstract This chapter deals with the modeling of Cu-based on-chip interconnects. The model considers the nonlinear effects of CMOS driver as well as the transmission line effects of interconnect line. The CMOS driver is represented by the nth power law model and the coupled-multiple interconnect lines are modeled by the FDTD technique. The model is validated by the industry standard HSPICE simulator. It is observed that the results of the proposed model closely match with that of HSPICE simulations. Encouragingly, the proposed model is highly time efficient than the HSPICE.

Keywords Complementary metal oxide semiconductor (CMOS) · Crosstalk · Finite-difference time-domain (FDTD) · Propagation delay · Transient response · Transmission line model

3.1 Introduction

In general, the modeling of on-chip interconnects is performed by assuming the nonlinear CMOS driver as a linear resistor [1–3]. However, this is not a valid assumption because the MOSFET operates in cutoff, linear, and saturation regions [4]. Moreover, the value of change in resistance in saturation region is much higher than the linear region. Especially, the PMOS operates in the saturation region for more than 60 % of time and in the linear region for less than 5 % of time [5]. Thus, assuming that the transistor operates in the linear region, leads to severe errors in the performance estimation of the driver-interconnect-load system.

The modeling of on-chip interconnects with nonlinear CMOS driver suffers with domain conversion problem. This problem arises because the CMOS driver elements appeared in the time domain, whereas the on-chip interconnects were solved in the frequency domain. The best way to avoid this conversion problem is the use of the FDTD technique to solve the transmission line equations of on-chip interconnects. Previously, Paul [6] analyzed the transmission line equations for the resistive driver and resistive load. In [7], Orlandi et al. extended this model by

© The Author(s) 2016

B.K. Kaushik et al., *Crosstalk in Modern On-Chip Interconnects*,
SpringerBriefs in Applied Sciences and Technology,
DOI 10.1007/978-981-10-0800-9_3

incorporating the frequency-dependent losses using the state-variable analysis. Additionally, several authors have also proposed FDTD-like techniques to analyze the transmission lines namely, latency insertion method [8], and alternating direction explicit-latency insertion method [9]. However, these models [6–9] analyze the transmission lines that are excited and terminated by the resistive drivers and resistive loads, respectively. Therefore, these models are not valid for the performance analysis of on-chip interconnects that are actually excited and terminated by the CMOS inverters.

In 2011, Li et al. [10] proposed a FDTD model for the analysis of on-chip interconnects by incorporating the nonlinear effects of CMOS driver. The authors analyzed the interconnect lines at global lengths for 180 nm technology node. However, they ignored the drain conductance parameter (σ), and therefore the estimated current was higher than the actual value. Moreover, the performance analysis is limited to functional crosstalk, wherein only one of the lines is in switching mode and the rest in quiet mode. In addition to the functional crosstalk analysis, the dynamic crosstalk also frequently occurs in the current technology nodes and its analysis is essential. The model presented in this chapter considers all these effects appropriately and accurately estimates the crosstalk-induced performance parameters.

This chapter presents the FDTD model for the crosstalk-induced performance analysis of on-chip interconnects. The crosstalk effects are comprehensively studied by including the dynamic and functional crosstalk analysis. The proposed model incorporates the transmission line effects of interconnect line and nonlinear effects of CMOS driver, and therefore it proves to be more accurate than the existing models. The CMOS driver is represented by the nth power law model [11], and the on-chip interconnect lines are modeled by FDTD technique.

This chapter is organized in seven sections as follows: Sect. 3.1 introduces the current research scenario on the modeling of on-chip interconnects. The motivation behind this work is described in Sect. 3.2. The FDTD model combined with nth power law model is presented in Sect. 3.3. The model is presented for two coupled interconnect lines; however, it can be extended to n coupled lines by changing the dimensions of the matrices. The model is validated with HSPICE in Sect. 3.4 for coupled-two lines and for coupled-three lines in Sect. 3.5. Finally, Sect. 3.6 presents the concluding remarks.

3.2 Motivation

In most of the analysis carried out, the CMOS gate drivers are approximately represented by a resistive drivers while performing the on-chip interconnect analysis [1–3]. The value of the equivalent resistance (R_{eq}) of driver is evaluated by averaging the values at the endpoints of the transition region. Using the Taylor expansion, R_{eq} is expressed as [4]

$$R_{eq} = \frac{3}{4} \frac{V_{DD}}{I_{DSAT}} \left(1 - \frac{5}{6} \sigma V_{DD} \right) \qquad (3.1)$$

where σ is the finite drain conductance parameter and I_{DSAT} is drain saturation current.

Figure 3.1 shows two different interconnect structures. First, Fig. 3.1 shows the interconnect lines driven by CMOS inverter and second, Fig. 3.2 shows the interconnect lines driven by resistive driver. In Fig. 3.1, R, C, and L represent per unit length (p.u.l.) line resistance, ground capacitance and line inductance, respectively, and C_{12} and L_{12} represent per unit length (p.u.l.) coupling capacitance and mutual inductance, respectively.

To justify the motivation, the transient response of Figs. 3.1 and 3.2 are compared using HSPICE. For symmetric driving capability of CMOS inverter, the width of PMOS is chosen as twice of NMOS width [4]. The time domain response is performed at 32 nm technology node at the global level interconnect length of 1 mm. The input transition time and supply voltage are considered as 10 ps and 0.9 V, respectively. The dimensions of the interconnect line are considered by the following two assumptions: (1) the space between the two interconnects is equal to the width of interconnect; and (2) the height from the ground plane is equal to the

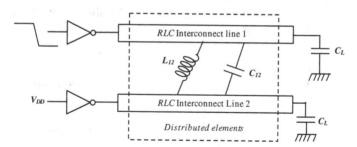

Fig. 3.1 CMOS gate-driven coupled interconnect lines

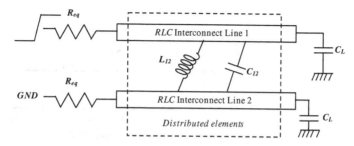

Fig. 3.2 Resistive gate-driven coupled interconnect lines

Table 3.1 Parasitic components of coupled interconnect lines

Line resistance R (kΩ/m)	Line capacitance C (pF/m)	Coupling capacitance C_{12} (pF/m)	Line inductance L (μH/m)	Mutual inductance L_{12} (μH/m)
150	15.11	98.59	1.645	1.484

thickness of the line. The resistivity of the copper material and the relative permittivity of the inter layer dielectric medium are chosen as 2.2 ($\mu\Omega$-cm) and 2.2, respectively. The interconnect line width and the aspect ratio are considered as 0.22 μm and 3, respectively [12]. For the above-mentioned interconnect dimensions, the associated parasitic values are listed in Table 3.1. The load capacitance C_L is considered as 2 fF.

The transient response comparison of on-chip interconnects using the CMOS driver and the resistive drivers are shown in Figs. 3.3, 3.4, and 3.5. For comprehensive analysis the functional, dynamic in-phase, dynamic out-phase crosstalk conditions are compared and shown in Figs. 3.3, 3.4 and 3.5, respectively. From these figures, a large deviation in the timing response can be observed between the resistive and CMOS gate-driven interconnects. For more clarification, the percentage error is compared while measuring the propagation delay during the dynamic out-phase switching. This error is observed to be as large as 68 %. This corroborates the earlier observations in [5, 10, 13], that the resistive driver model presents inaccurate results for the on-chip interconnect performance.

Fig. 3.3 Functional crosstalk transient response of CMOS and resistive drivers

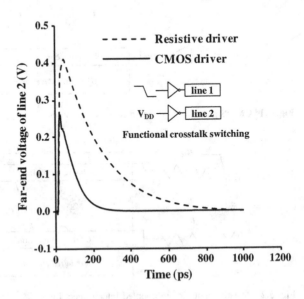

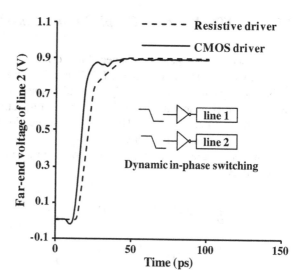

Fig. 3.4 Dynamic in-phase crosstalk transient response of CMOS and resistive drivers

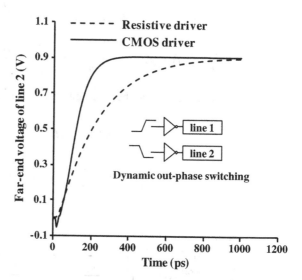

Fig. 3.5 Dynamic out-phase crosstalk transient response of CMOS and resistive drivers

3.3 FDTD Model of CMOS Gate-Driven Cu Interconnects

In this section, the FDTD model is presented for the performance analysis of two coupled interconnects. In a more accurate manner, the nonlinear CMOS driver effects are incorporated in the proposed model using the nth power law model. Moreover, the short-channel effects including the velocity saturation and finite drain conductance parameter are also considered in the proposed model.

3.3.1 FDTD Model of Coupled Interconnects

The coupled-two copper interconnect lines driven by CMOS driver is shown in Fig. 3.6, where R is line resistance, L is line inductance, C is line capacitance, and C_L is load capacitance, respectively. The subscripts 1 and 2 represent the parasitic values to the corresponding line 1 and line 2, respectively. All the line parasitic values are mentioned in p.u.l. The time and space along the interconnect line are represented by t and z, respectively.

Using the telegrapher's equations the coupled transmission line equations can be expressed as

$$\frac{\partial}{\partial z} V_1(z,t) + L_1 \frac{\partial}{\partial t} I_1(z,t) + L_{12} \frac{\partial}{\partial t} I_2(z,t) + R_1 I_1(z,t) = 0 \tag{3.2a}$$

$$\frac{\partial}{\partial z} V_2(z,t) + L_2 \frac{\partial}{\partial t} I_2(z,t) + L_{12} \frac{\partial}{\partial t} I_1(z,t) + R_2 I_2(z,t) = 0 \tag{3.2b}$$

$$\frac{\partial}{\partial z} I_1(z,t) + (C_1 + C_{12}) \frac{\partial}{\partial t} V_1(z,t) - C_{12} \frac{\partial}{\partial t} V_2(z,t) = 0 \tag{3.2c}$$

$$\frac{\partial}{\partial z} I_2(z,t) + (C_2 + C_{12}) \frac{\partial}{\partial t} V_2(z,t) - C_{12} \frac{\partial}{\partial t} V_1(z,t) = 0 \tag{3.2d}$$

Equations (3.2a)–(3.2d) can be represented in the matrix form as

$$\frac{d}{dz} V(z,t) + RI(z,t) + L \frac{d}{dt} I(z,t) = 0 \tag{3.2e}$$

$$\frac{d}{dz} I(z,t) + \frac{d}{dt} CV(z,t) = 0 \tag{3.2f}$$

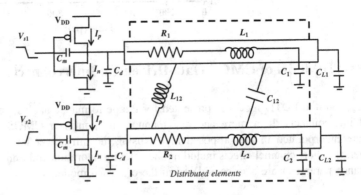

Fig. 3.6 Coupled-two interconnect lines driven by CMOS inverter

where the voltages and currents are expressed in a 2×1 column vectors and the line parasitic are expressed in 2×2 matrices as shown below

$$V = \begin{bmatrix} V_1 \\ V_2 \end{bmatrix}, \; I = \begin{bmatrix} I_1 \\ I_2 \end{bmatrix}, \; R = \begin{bmatrix} R_1 & 0 \\ 0 & R_2 \end{bmatrix}, \; L = \begin{bmatrix} L_1 & L_{12} \\ L_{12} & L_2 \end{bmatrix} \text{ and}$$

$$C = \begin{bmatrix} C_1 + C_{12} & -C_{12} \\ -C_{12} & C_2 + C_{12} \end{bmatrix}.$$

Equations (3.2e) and (3.2f) can be solved using the central difference approximation method. However, the results shows better accuracy if the V and I points are separated in space location by half of the space discretization, i.e., $\Delta z/2$. In a similar manner, the V and I points are separated in time location by half of the time discretization, i.e., $\Delta t/2$. This information can also be visualized in Fig. 3.7.

The interconnect line is driven by the nonlinear CMOS driver at the near-end boundary ($z = 0$) and terminated by a capacitive load at the far-end boundary ($z = l$). The discretized solution points of V and I along the line are shown in Fig. 3.8, where the Nz represents the number of space segments that can be derived from $Nz = l/\Delta z$.

Applying the finite-difference approximations from Fig. 3.7, Eqs. (3.2e) and (3.2f) can be solved as

$$\frac{V_{k+1}^{n+1} - V_k^{n+1}}{\Delta z} + L \frac{I_k^{n+3/2} - I_k^{n+1/2}}{\Delta t} + R \frac{I_k^{n+3/2} + I_k^{n+1/2}}{2} = 0 \qquad (3.3a)$$

$$I_k^{n+3/2} = BDI_k^{n+1/2} + B\left(V_k^{n+1} - V_{k+1}^{n+1}\right) \text{ for } k = 1, 2, \ldots, Nz \qquad (3.3b)$$

where $B = \left[\frac{\Delta z}{\Delta t}L + \frac{\Delta z}{2}R\right]^{-1}, D = \left[\frac{\Delta z}{\Delta t}L - \frac{\Delta z}{2}R\right]$

$$\frac{I_k^{n+1/2} - I_{k-1}^{n+1/2}}{\Delta z} + C \frac{V_k^{n+1} - V_k^n}{\Delta t} = 0 \qquad (3.4a)$$

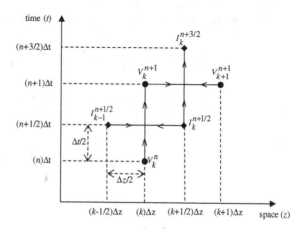

Fig. 3.7 Spatial and time discretization of FDTD technique to achieve second order accuracy

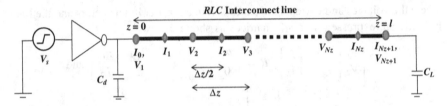

Fig. 3.8 Space discretization of FDTD technique on the interconnect line

$$V_k^{n+1} = V_k^n + A\left(I_{k-1}^{n+1/2} - I_k^{n+1/2}\right) \quad \text{for } k = 2, 3, \ldots\ldots, Nz \qquad (3.4b)$$

where $A = \left[\frac{\Delta z}{\Delta t} C\right]^{-1}$.

Here, it can be noticed that the calculations are interleaved in both space and time. For example, in (3.3b) the updated value of I is calculated from the past value of I and the most recent values of V. The vectors of V and I are represented as

$$V_i^j = V[i\Delta z, j\Delta t] , \quad I_i^j = I[(i+1/2)\Delta z, j\Delta t] \qquad (3.5)$$

3.3.2 Incorporation of Boundary Constraints

This section discusses the incorporation of nonlinear CMOS driver effects in the FDTD model. At near-end boundary, the voltage and current points are denoted by V_1 and I_0, respectively. From Fig. 3.8, it can be noted that the space discretization between the current values I_0 and I_1 is $\Delta z/2$. Therefore, at $k = 1$ Eq. (3.4b) becomes

$$V_1^{n+1} = V_1^n + 2A\left[I_0^{n+1/2} - I_1^{n+1/2}\right] \qquad (3.6a)$$

In Eq. (3.6a), the source current I_0 at $(n + (1/2))$ time interval is obtained by averaging the values at n and $n + 1$, then the Eq. (3.6a) becomes

$$V_1^{n+1} = V_1^n + 2A\left[\frac{I_0^{n+1} + I_0^n}{2} - I_1^{n+1/2}\right] \qquad (3.6b)$$

here, I_0 represents the driving current of the CMOS. Using the KCL at near-end boundary, the driving current can be expressed as

$$I_0 = C_m\left[\frac{d(V_s - V_1)}{dt}\right] + I_p - I_n - C_d\frac{dV_1}{dt} \qquad (3.7)$$

where C_m and C_d are the drain to gate coupling capacitance and drain diffusion capacitance, respectively. The PMOS and NMOS currents are represented by I_p and I_n, respectively. These currents can be expressed using the nth power law model as [11]

$$
I_p = \begin{cases} 0 & V_S \geq V_{DD} - |V_{Tp}| & \text{(cutoff)} \\ I_{DSATp}(1 + \sigma_p(V_{DD}-V_1))\left(2 - \frac{V_{DD}-V_1}{V_{DSATp}}\right)\frac{V_{DD}-V_1}{V_{DSATp}} & V_1 > V_{DD} - V_{DSATp} & \text{(lin)} \\ I_{DSATp}(1 + \sigma_p(V_{DD}-V_1)) & V_1 \leq V_{DD} - V_{DSATp} & \text{(sat)} \end{cases}
$$

$$(3.8a)$$

$$
I_n = \begin{cases} 0 & V_S \leq V_{Tn} & \text{(cutoff)} \\ I_{DSATn}(1 + \sigma_n V_1)\left(2 - \frac{V_1}{V_{DSATn}}\right)\frac{V_1}{V_{DSATn}} & V_1 < V_{DSATn} & \text{(lin)} \\ I_{DSATn}(1 + \sigma_n V_1) & V_1 \geq V_{DSATn} & \text{(sat)} \end{cases}
$$

$$(3.8b)$$

where V_{DSAT}, V_T, I_{DSAT}, and σ are the drain saturation voltage, drain saturation current, threshold voltage, finite drain conductance parameter. The subscripts of p and n represent the PMOS and NMOS, respectively. The V_{DSAT} and I_{DSAT} are expressed as

$$V_{DSATp} = K_p(V_{DD} - V_S - |V_{Tp}|)^{m_p} \qquad (3.8c)$$

$$V_{DSATn} = K_n(V_S - V_{Tn})^{m_n} \qquad (3.8d)$$

$$I_{DSATp} = \frac{W_p}{L_{eff}} B_p(V_{DD} - V_S - |V_{Tp}|)^{s_p} \qquad (3.8e)$$

$$I_{DSATn} = \frac{W_n}{L_{eff}} B_n(V_S - V_{Tn})^{s_n} \qquad (3.8f)$$

The parameters s and B control the saturation region while m and K control the linear region characteristics. The width and effective channel length of MOSFET is represented by W and L_{eff}, respectively. Using [11], the model parameters can be computed and the values are listed in Table 3.2.

Table 3.2 MOS parameters at 32 nm technology node

Parameter	NMOS	PMOS
m	0.211	0.087
s	0.915	1.07
B	3.55×10^{-5}	0.801×10^{-5}
K	0.369	0.316
σ	0.867	3.11
V_T	0.36	0.366

The source current I_0 can be discretized as:

$$I_0^{n+1} = C_m \frac{V_s^{n+1} - V_s^n}{\Delta t} + I_p^{n+1} - I_n^{n+1} - (C_m + C_d)\frac{V_1^{n+1} - V_1^n}{\Delta t} \quad (3.9)$$

Using (3.9) and (3.6b)

$$V_1^{n+1} = V_1^n + EA\left[\frac{C_m}{\Delta t}(V_s^{n+1} - V_s^n) + I_0^n\right] - 2EAI_1^{n+1/2} + EA\left(I_p^{n+1} - I_n^{n+1}\right) \quad (3.10)$$

where $E = \left[U + \frac{A}{\Delta t}(C_m + C_d)\right]^{-1}$ and U is identity matrix.

At the far-end boundary, the voltage and current equations can be derived as follows:

Using Eq. (3.4b), the output voltage becomes

$$V_{Nz+1}^{n+1} = V_{Nz+1}^n + 2A\left(I_{Nz}^{n+1/2} - \frac{I_{Nz+1}^{n+1} + I_{Nz+1}^n}{2}\right) \quad (3.11)$$

For a capacitive load C_L, the output current is expressed as

$$I_{Nz+1} = C_L \frac{d}{dt}V_{Nz+1} \quad (3.12a)$$

Equation (3.12a) is discretized as

$$I_{Nz+1}^{n+1} = C_L \frac{(V_{Nz+1}^{n+1} - V_{Nz+1}^n)}{\Delta t} \quad (3.12b)$$

Using (3.11) and (3.12b)

$$V_{Nz+1}^{n+1} = V_{Nz+1}^n + 2FA\left[I_{Nz}^{n+1/2} - \frac{I_{Nz+1}^n}{2}\right] \quad (3.13)$$

where $F = \left[U + \frac{AC_L}{\Delta t}\right]^{-1}$.

The voltage and current expressions are evaluated in a bootstrapping fashion. First, the voltages are evaluated at a fixed time using Eqs. (3.10), (3.4b), and (3.13). Second, the currents are evaluated from (3.9), (3.3b), and (3.13). However, in order to get the stable output, the maximum value of time step must be less than $\Delta z/v$. It is worth noting that the boundary conditions are implicitly derived and interconnect line equations are explicitly derived. Therefore, there is no stability issue at the boundaries and stability of the driver-interconnect-line system is completely determined by the interconnect line. The proposed model is derived for the two coupled interconnect lines. For extended coupled interconnect lines, these equations remain valid except the changes in the matrix dimensions. For instance, to model the coupled-three interconnect lines, the interconnect parasitic values have to be mentioned in 3×3 order and voltage and current values should be derived in 3×1 order.

3.4 Validation of the Model

The two coupled interconnect lines are considered for the validation of the proposed model. The results of the proposed are compared against the industry HSPICE simulation results. Using the similar simulation environment as described in Sect. 3.2, the following interconnect parasitic values are used in the simulations

$$R = \begin{bmatrix} 150 & 0 \\ 0 & 150 \end{bmatrix} \frac{k\Omega}{m}, \quad L = \begin{bmatrix} 1.645 & 1.484 \\ 1.484 & 1.645 \end{bmatrix} \frac{\mu H}{m}, \text{ and}$$

$$C = \begin{bmatrix} 113.7 & -98.59 \\ -98.59 & 113.7 \end{bmatrix} \frac{pF}{m}$$

For the above parasitic values, the signal mode velocities of a lossless interconnect line are $v_{m1} = 1.4 \times 10^8$ m/s and $v_{m2} = 1.7 \times 10^8$ m/s. Using the break frequency of 3.1×10^{10} Hz, the minimum space discretization is calculated as 4.5×10^{-4} m. Based on the CFL condition, the time discretization is obtained to be less than 2.67×10^{-12} s, for a larger mode velocity, v_{m2} [6]. The rise and fall transition of the input signal is assumed to be 10 ps. In the two coupled interconnect lines, line 2 is considered as victim line and accordingly all the performance parameters such as propagation delay, noise peak voltage and its timing instances are evaluated on the victim line 2.

The functional crosstalk analysis is studied by switching the aggressor line 1 and keeping the victim line 2 in quiescent mode [14]. Later on, dynamic crosstalk analysis is studied by switching both aggressor and victim lines either in in-phase or out-phase. In all these cases, the transient responses at far-end of the victim line are compared and shown in Figs. 3.9, 3.10 and 3.11. From these figures, it can be observed that the proposed model accurately captures the timing response in all switching cases, whereas the model presented in [10] is unable to capture the response accurately. This is due to the fact that the model presented in [10] ignores the drain conductance parameter and the current in saturation region is assumed to be independent of drain voltage.

To test the robustness of the proposed model, the percentage errors are calculated while measuring the crosstalk-induced peak voltage and its timing instances [15]. The percentage error is calculated with respect to the HSPICE simulation results. For comparison with the existing models, the model in [10] is also considered as a reference. The peak voltage values and timing instants are shown in Tables 3.3 and 3.4, respectively. From these tables, it has been observed that the average errors using proposed model and model in [10] are 1.5 and 14 %, respectively with respect to the HSPICE simulations.

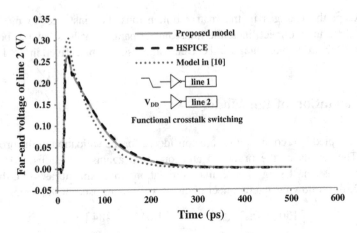

Fig. 3.9 Comparison of functional crosstalk transient response of line 2

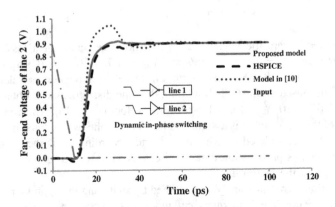

Fig. 3.10 Comparison of dynamic in-phase transient response of line 2

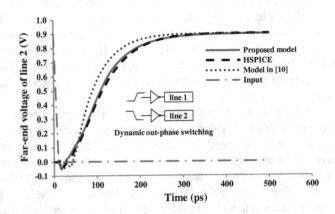

Fig. 3.11 Comparison of dynamic out-phase transient response of line 2

Table 3.3 Comparison of computational error involved in peak noise voltage using HSPICE and proposed model

Input transition time (T_r) (ps)	Peak voltage				
	Proposed model (V)	Li et al. model [10] (V)	HSPICE (V)	% error proposed model	% error [10]
10	0.26	0.30	0.26	0.0	−15.3
30	0.25	0.29	0.24	−4.1	−20.8
50	0.23	0.27	0.22	−4.5	−22.7
70	0.21	0.24	0.21	0.0	−14.2
90	0.20	0.22	0.2	0.0	−10.0

Table 3.4 Computational of computational error involved in peak voltage timing using HSPICE and proposed model

Input transition time (T_r) (ps)	Peak voltage timing				
	Proposed model (ps)	Li et al. model [10] (ps)	HSPICE (ps)	% error propoposed model	% error [10]
10	25.90	26.40	25.74	−0.6	−2.5
30	39.30	28.80	40.89	3.8	29.5
50	59.10	58.40	60.39	2.1	3.3
70	78.80	75.70	79.39	0.7	4.6
90	98.70	90.10	98.89	0.1	8.8

The propagation delay is also compared between the proposed model and the model presented in [10]. Figure 3.12 shows the propagation delay comparison for different values of input transition timings. From Fig. 3.12, it can be observed that the delay during the out-phase transition is higher than the in-phase transition delay.

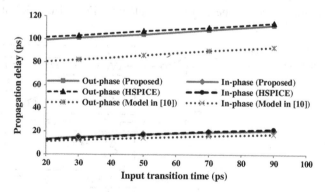

Fig. 3.12 Propagation delay variation with transition time

It is due to the effect of Miller capacitance. Moreover, it has been observed that the proposed model captures the delay accurately during both in-phase and out-phase transitions. The average error is observed to be less than 3 %, whereas using [10] the average error is as high as 17 %.

3.5 Extended Three Coupled Interconnect Lines

In this section, the three coupled interconnect lines are analyzed using the proposed model. The three coupled interconnects that are driven by the CMOS driver is shown in the Fig. 3.13. Similar to the two coupled interconnect lines, the input rise and fall transitions are considered as 10 ps and load capacitance is considered as 2 fF. The following interconnect parasitic values are used in the simulations:

$$
R = \begin{bmatrix} 150 & 0 & 0 \\ 0 & 150 & 0 \\ 0 & 0 & 150 \end{bmatrix} \frac{k\Omega}{m}, \ L = \begin{bmatrix} 1.645 & 1.484 & 1.264 \\ 1.484 & 1.645 & 1.484 \\ 1.264 & 1.484 & 1.645 \end{bmatrix} \frac{\mu H}{m} \ \text{and}
$$

$$
C = \begin{bmatrix} 113.7 & -98.59 & 0 \\ -98.59 & 212.29 & -98.59 \\ 0 & -98.59 & 113.7 \end{bmatrix} \frac{pF}{m}
$$

From the values of capacitance, the coupling capacitance between the lines 1 and 3, C_{13} can be safely neglected because of the shielding of line 2 [3]. In the three coupled interconnect line system, line 1 and line 3 are considered as aggressor lines and line 2 is considered as victim line. The transient responses at the far-end terminal of the victim line are compared at different switching cases and shown in Fig. 3.14. It can be observed that in all switching cases, the proposed model provides the timing responses as accurate as that of HSPICE.

The percentage error involved while measuring the propagation delay on victim line 2 at different switching cases is shown in Table 3.5. The percentage error is calculated with reference to the HSPICE simulations. From Table 3.5, it has been

Fig. 3.13 CMOS gate-driven three coupled interconnect lines

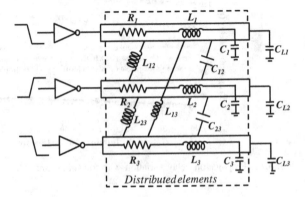

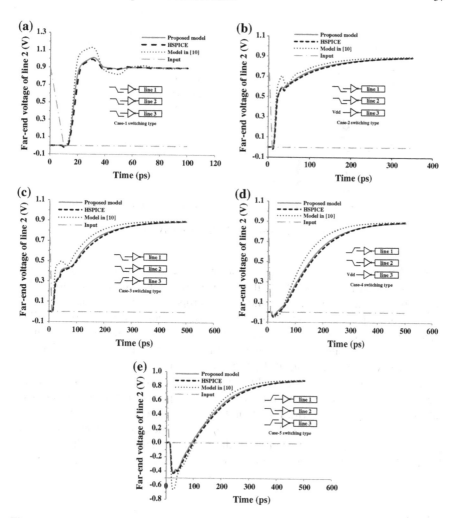

Fig. 3.14 Comparison of transient response on victim line 2 during **a** case-1, **b** case-2, **c** case-3, **d** case-4, and **e** case-5 input switching modes

Table 3.5 Comparison of percentage error involved for propagation delay on victim line 2

Input switching modes				Propagation delay on victim line 2				
Case-mode	Line 1 (agg.)	Line 2 (vic.)	Line 3 (agg.)	Proposed model (ps)	Li et al. model [10] (ps)	HSPICE (ps)	% error proposed model	% error [10]
Case-1	$1 \rightarrow 0$	$1 \rightarrow 0$	$1 \rightarrow 0$	12.1	11.5	12.7	4.7	9.4
Case-2	$1 \rightarrow 0$	$1 \rightarrow 0$	V_{DD}	15	13.8	15.7	4.4	12.1
Case-3	$1 \rightarrow 0$	$1 \rightarrow 0$	$0 \rightarrow 1$	78.1	20.9	79.6	1.8	73.7
Case-4	$0 \rightarrow 1$	$1 \rightarrow 0$	V_{DD}	133	111	138.4	3.9	19.8
Case-5	$0 \rightarrow 1$	$1 \rightarrow 0$	$0 \rightarrow 1$	174.9	164.7	181.4	3.5	9.2

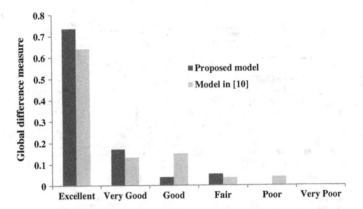

Fig. 3.15 Histogram forms the datasets of Fig. 3.14a using FSV tool

observed that the average error using the proposed model is 4 % and using the model presented in [10] is 24 %. It is also observed that the propagation delay increases with increasing the case mode. The propagation delay changes by more than one order from case-1 mode to case-5 mode switching. It is due to the Miller capacitance that highly influences the effective capacitance value when two lines are switching oppositely. Accordingly, for three coupled lines the case-5 is the worst case switching scenario, if the line 2 is considered as a victim.

Lastly, in order to make a quality of the comparison the data values of the Fig. 3.14a are changed to the natural language descriptor using the feature selective validation (FSV) tool [16–18]. Based on the global difference measure (GDM), the quality of the comparison can be of different types such as excellent, very good, good, fair, poor, and very poor. The result of the FSV tool is shown in Fig. 3.15. The result describes that the proposed model is as accurate as that of HSPICE.

3.6 Summary

This chapter presented an accurate FDTD model for the crosstalk-induced performance analysis of on-chip interconnects. In a more realistic manner, the nonlinear CMOS driver effects are incorporated in the proposed model. The finite drain conduction parameter is also incorporated to improve the accuracy. The boundary conditions at the interface are derived in an implicit manner and therefore the stability of the FDTD solution is strictly followed by the Courant condition. The results of the proposed model are compared with the HSPICE simulations. The comparison results show that the model captures the timing response, propagation delay, and peak voltage quite well. The proposed model is applicable to asymmetric drivers and line configurations as well.

References

1. Agarwal K, Sylvester D, Blaauw D (2006) Modeling and analysis of crosstalk noise in coupled RLC interconnects. IEEE Trans Comput Aided Des Integr Circ Syst 25(5):892–901
2. Edelstein D, Heidenreich J, Goldblatt R, Cote W, Uzoh C, Lustig N, Roper P, McDevitt T, Motsiff W, Simon A, Dukovic J, Wachnik R, Rathore H, Schulz R, Su L, Luce S, Slattery J (1997) Full copper wiring in a sub-0.25 μm CMOS ULSI technology. In: Proceedings of International Electron Devices Meeting Techincal Digest, USA, pp 773–776
3. Davis JA, Meindl JD (2000) Compact distributed RLC models, part II: coupled line transient expressions and peak crosstalk in multilevel networks. IEEE Trans Electron Devices 47(11):2078–2087
4. Rabaey JM, Chandrakasan A, Nikolic B (2003) Digital integrated circuits: a design perspective, 2nd edn. Prentice-Hall
5. Kaushik BK, Sarkar S (2008) Crosstalk analysis for a CMOS-gate-driven coupled interconnects. IEEE Trans Comput Aided Des Integr Circ Syst 27(6):1150–1154
6. Paul CR (1994) Incorporation of terminal constraints in the FDTD analysis of transmission lines. IEEE Trans Electromagn Compat 36(2):85–91
7. Orlandi A, Paul CR (1996) FDTD analysis of lossy, multiconductor transmission lines terminated in arbitrary loads. IEEE Trans Electromagn Compat 38(3):388–399
8. Schutt-ainé JE (2001) Latency insertion method (LIM) for the fast transient simulation of large networks. IEEE Trans Circ Syst I Fundam Theory Appl 48(1):81–89
9. Kurobe H, Sekine T, Asai H (2012) Alternating direction explicit-latency insertion method (ADE-LIM) for the fast transient simulation of transmission lines. IEEE Trans Compon Packag Manuf Technol 2(5):783–792
10. Li XC, Ma JF, Swaminathan M (2011) Transient analysis of CMOS gate driven RLGC interconnects based on FDTD. IEEE Trans Comput Aided Des Integr Circ Syst 30(4):574–583
11. Sakurai T, Newton AR (1991) A simple MOSFET model for circuit analysis. IEEE Trans Electron Devices 38(4):887–894
12. International Technology Roadmap for Semiconductors (2013) http://public.itrs.net
13. Kaushik BK, Sarkar S, Agarwal RP, Joshi RC (2010) An analytical approach to dynamic crosstalk in coupled interconnects. Microelectron J 41(2):85–92
14. Krishnamurthy R, Sharma GK (2013) An area efficient wide range on-chip delay measurement architecture. In: Proceedings of Springer VLSI Design and Test (VDAT 2013), Jaipur, pp 49–58
15. Kumar VR, Kaushik BK, Patnaik A (2014) An accurate FDTD model for crosstalk analysis of CMOS-gate-driven coupled RLC interconnects. IEEE Trans Electromagn Compat 56(5):1185–1193
16. IEEE Standard P1597 (2008) Standard for validation of computational electromagnetics computer modeling and simulation—Part 1, 2
17. Duffy AP, Martin AJM, Orlandi A, Antonini G, Benson TM, Woolfson MS (2006) Feature selective validation (FSV) for validation of computational electromagnetics (CEM). Part I— the FSV method. IEEE Trans Electromagn Compat 48(3):449–459
18. Orlandi A, Duffy AP, Archambeault B, Antonini G, Coleby DE, Connor S (2006) Feature selective validation (FSV) for validation of computational electromagnetics (CEM). Part II— assessment of FSV performance. IEEE Stand 48(3):460–467

Chapter 4
FDTD Model for Crosstalk Analysis of Multiwall Carbon Nanotube (MWCNT) Interconnects

Abstract This chapter introduces an equivalent single conductor (ESC) model of MWCNT interconnects. Based on the ESC model, this chapter presents an accurate FDTD model of MWCNT while incorporating the quantum effects of nanowire and nonlinear effects of CMOS driver. To reduce the computational effort required for analyzing the CMOS driver, a simplified but accurate model is employed named as modified alpha-power law model.

Keywords Crosstalk · Equivalent *RLC* model · Kinetic inductance · Multiwall carbon nanotube (MWCNT) quantum resistance · Quantum capacitance

4.1 Introduction

The conventional interconnect copper material is unable to meet the requirements of future technology needs, since it suffers from low reliability with downscaling of interconnect dimensions. Moreover, the resistivity of copper increases, due to electron-surface scattering and grain boundary scattering with smaller dimensions. Therefore, researchers are forced to find an alternative material for global VLSI interconnects. Carbon nanotubes have been proposed to be one of the potential candidates for VLSI interconnects due to their unique physical properties, such as extraordinary mobility, large mean free path, and high current carrying capability [1, 2].

Carbon nanotubes can be classified into single-walled carbon nanotube (SWCNT) and multiwalled carbon nanotube (MWCNT) [3–6]. The promising interconnect solution for global interconnect lengths are MWCNTs due to their high current carrying capabilities than SWCNT bundles. Naeemi et al. observed that for longer interconnects, MWCNTs can have conductivities several times greater than SWCNT bundles [6]. Hence, many researchers consider the MWCNTs as a potential solution for global interconnect material. The experimental and theoretical investigations of MWCNTs as interconnect material have been presented in [7] and [8], respectively.

The performance of an MWCNT interconnect line is generally evaluated by means of an equivalent transmission line model. Li et al. proposed a multiconductor transmission line (MTL) model to represent the MWCNT interconnect [9]. However, the analysis of MWCNT using the MTL model can be computationally expensive. For this reason, the equivalent single conductor (ESC) model was proposed in [8], using the assumption that voltage at an arbitrary cross section along MWCNT are the same, such that all nanotubes are connected in parallel at the both ends. The accuracy of the ESC model has been verified by several researchers [4, 8, 10]. They observed that the transient responses of ESC model and MTL model are in good agreement.

The FDTD technique has been used widely to analyze the transmission lines due to their better accuracy [11]. However, incorporation of different boundary conditions in the FDTD models is a challenging task. Previously, Paul [12, 13] incorporated the boundary conditions to analyze the transmission lines for resistive driver and resistive load boundaries. However, these studies were focused only on copper interconnects and hence, not suitable for next-generation graphene-based nanointerconnects. The quantum and contact resistances at the near-end and far-end terminals of a nanointerconnect line results in complex boundary conditions. For the first time, Liang et al. [14] proposed a crosstalk noise model for the analysis of MWCNT interconnects using FDTD technique. However, the authors represented the nonlinear CMOS driver by a resistive driver, thus limiting the accuracy of their model. Moreover, they did not validate their proposed model with respect to HSPICE. Therefore, a more accurate model is required that allows a better crosstalk-induced performance estimation of MWCNT interconnects.

The fabrication technique of MWCNT bundles was reported in [15], using thermal chemical vapor deposition technique. The authors have demonstrated the feasibility of growing perfectly aligned carbon nanotube bundles. Recently, Wang et al. [16] fabricated the MWCNTs arrays using microwave plasma chemical vapor deposition on Si substrate with interdigital electrodes. This method is able to control the thickness of MWCNT arrays based on the growth time. Although, the controlled growth of MWCNTs with high CNT density is realizable, the researchers are still facing some challenges in terms of large imperfect metal–nanotube contact resistance, poor control on number of shells, chirality and orientation, higher growth temperature during the fabrication process. However, efforts are underway to fabricate MWCNTs for interconnect applications.

This chapter presents an accurate numerical model for comprehensive crosstalk analysis of coupled MWCNT interconnects based on FDTD method. Using this method, the voltage and current can be accurately estimated at any particular point on the interconnect line. Since the proposed model requires less number of assumptions, the accuracy is very high. The nonlinear CMOS driver effects are incorporated using the modified alpha-power law model with suitable boundary conditions. Using the proposed FDTD method, the functional and dynamic crosstalk analysis is carried out. The results demonstrate that the proposed model has high accuracy that matches closely with the HSPICE results. In addition to this, the proposed model is highly time efficient than the HSPICE. Although, this chapter demonstrates the crosstalk effects on two coupled interconnect lines, the model can be extended to N lines.

The rest of the chapter is organized as follows: Sect. 4.2 describes the ESC model of an MWCNT. In Sect. 4.3, the FDTD method is developed for coupled MWCNT interconnect lines. Section 4.4 is devoted to the validation of proposed model for coupled-two lines. In Sect. 4.5, the sensitivity analysis is performed to evaluate the validity of the assumptions associated with the proposed model. Finally, Sect. 4.6 concludes this chapter.

4.2 Equivalent Single Conductor Model of the MWCNT Interconnect

This section presents an equivalent RLC model of an MWCNT interconnect line. Consider a horizontal MWCNT bundle interconnect line positioned over a ground plane at a distance H and placed in a dielectric medium with dielectric constant ε. The geometry of an MWCNT interconnect is shown in Fig. 4.1. The coupling parasitics between the two MWCNT interconnects is shown in Fig. 4.2, where s is the spacing between the interconnect lines, and l_{12} and c_{12} represent the mutual inductance and coupling capacitance between the interconnect lines, respectively. The MWCNT interconnect consists of N number of tubes

$$N = 1 + \mathrm{int}\left[\frac{(d_N - d_1)}{2\delta}\right] \tag{4.1}$$

where δ, d_1, and d_N represent intershell distance, innermost shell diameter, and outermost shell diameter, respectively.

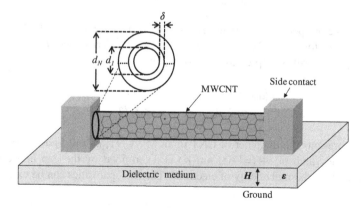

Fig. 4.1 Geometry of an MWCNT interconnect above the ground plane

Fig. 4.2 Cross-sectional view and coupling parasitics between the MWCNT interconnects

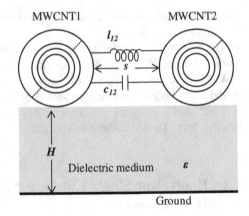

Fig. 4.3 Electrical equivalent model of an MWCNT interconnect

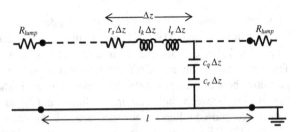

The MWCNT interconnect has been represented by an equivalent single conductor (ESC) model as shown in Fig. 4.3 [8]. The *RLC* parasitics of an MWCNT interconnect are primarily dependent on the number of conducting channels. The number of conducting channels in a CNT can be derived by adding all the subbands contributing to the current conduction. Using Fermi function, it can be expressed as

$$N_{\text{ch},i} = \sum_{\text{subbands}} \frac{1}{\exp(|E_i - E_\text{F}|/k_\text{B}\,T) + 1} \tag{4.2a}$$

where T is the temperature, k_B is the Boltzmann constant, and E_i is the lowest (or highest) energy for the subbands above (or below) the Fermi level E_F.

A simplified form of expression (4.2a) is [6]:

$$\begin{aligned} N_{\text{ch},i} &\approx k_1 T d_i + k_2 && d_i > d_\text{T}/T \\ &\approx 2/3 && d_i < d_\text{T}/T \end{aligned} \tag{4.2b}$$

where d_i represents the diameter of CNT in an MWCNT, k_1 and k_2 are curve fitted constants. The value of d_T (=1300 nm K) is determined by the gap between the subbands and the thermal energy of electrons. The *RLC* parasitics can be extracted a follows:

4.2.1 Resistance

Each shell in the MWCNT primarily demonstrates three different types of resistances: (1) quantum resistance (R_Q) due to the finite conductance value of quantum wire if there is no scattering along the length; (2) imperfect metal–nanotube contact resistance (R_{MC}) that exhibits a value ranging from zero to few kilo-ohms depending on the fabrication process [17–19]; and (3) scattering resistance (r_s) due to acoustic phonon scattering and optical phonon scattering that occurs when the nanotube lengths exceed the mean free path of electrons. The scattering resistance appeared as per unit length distributed resistance along the line, whereas (1) and (2) are considered as lumped resistances placed at the contacts of near-end and far-end terminals. The overall effective lumped resistance at the near-end/far-end terminals of the MWCNT can be expressed as

$$R_{\text{lump,ESC}} = \frac{1}{2}\left[\sum_{i=1}^{N}\left(\frac{R_Q}{2N_{\text{ch},i}} + R_{\text{MC},i}\right)^{-1}\right]^{-1} \quad \text{where} \quad R_Q = \frac{h}{e^2} \approx 25.8 \text{ K}\Omega$$

(4.3a)

The p.u.l. scattering resistance of an MWCNT can be expressed as

$$r_{\text{s,ESC}} = \frac{h/e^2}{\sum_{i=1}^{N} 2N_{\text{ch},i}\,\lambda_{\text{mfp},i}} \quad \text{where} \quad \lambda_{\text{mfp},i} = \frac{10^3 d_i}{(T/T_0) - 2}, \quad T_0 = 100 \text{ K} \quad (4.3b)$$

where h and e represent the Planck's constant and the charge of an electron, respectively.

4.2.2 Inductance

The MWCNT demonstrates two different types of inductances:

(1) Magnetic inductance: The magnetic inductance (l_e) is due to the magnetic field generation around a current-carrying conductor. In the presence of ground plane, the p.u.l. magnetic inductance of a CNT shell shown in Fig. 4.4 is given by [20]

$$l_e = \frac{\mu}{2\pi}\cosh^{-1}\left(\frac{d+2H}{d}\right)$$

(4.4a)

where d and H represent the shell diameter and height from the ground plane, respectively. Additionally, the intershell mutual inductance (l_m) is mainly due to the magnetic field coupling between the adjacent shells in an MWCNT. The p.u.l. l_m can be expressed as [9]

Fig. 4.4 A single CNT shell
above a ground plane

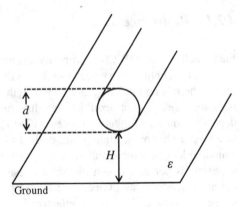

$$l_m = \frac{\mu}{2\pi} \ln\left(\frac{d_i}{d_{i-1}}\right) \tag{4.4b}$$

(2) Kinetic inductance: The kinetic inductance (l_k) is mainly due to the kinetic
   energy of electrons. By equating kinetic energy stored in each conducting
   channel of a CNT shell to the effective inductance, the kinetic inductance of
   each conducting channel (l'_k) in a CNT can be expressed as [20]

$$l'_k = \frac{h}{2e^2 v_F} \tag{4.4c}$$

where v_F is the Fermi velocity $\approx 8 \times 10^5$ m/s [21].

By adopting a recursive approach proposed in [8], the equivalent inductance
($l_{k,ESC}$) of Fig. 4.3 can be expressed as

$$l_{equ,1} = l_{k,1} \tag{4.4d}$$

$$l_{equ,i} = \left(\frac{1}{l_{equ,i-1} + l_m^{i-1,i}} + \frac{1}{l_{k,i}}\right)^{-1}, \quad i = 2, 3, \ldots, N \tag{4.4e}$$

$$l_{k,ESC} = l_{equ,N} \tag{4.4f}$$

where

$$l_m^{i-1,i} = \frac{\mu}{2\pi} \ln(d_i/d_{i-1}), \quad i = 2, 3, \ldots, N \tag{4.4g}$$

$$l_{k,i} = \frac{1}{2N_{ch,i}} \frac{h}{2e^2 v_F} \quad 1 \leq i \leq N \tag{4.4h}$$

4.2.3 Capacitance

The MWCNT interconnect consists of two types of capacitances:

(1) Electrostatic capacitance: It represents the electrostatic field coupling between the CNT and the ground plane. The electrostatic capacitance (c_e) of MWCNT appears between the external shell and the ground plane, as external shell shields the internal ones. The p.u.l. c_e of a CNT shell shown in Fig. 4.4 can be expressed as [20]

$$c_e = \frac{2\pi\varepsilon}{\cosh^{-1}\left(\frac{d+2H}{d}\right)} \tag{4.5a}$$

Additionally, the intershell coupling capacitance (c_m) is mainly due to the potential difference between adjacent shells in MWCNT. The p.u.l. c_m can be expressed as [9]

$$c_m = \frac{2\pi\varepsilon}{\ln\left(\frac{d_i}{d_{i-1}}\right)} \tag{4.5b}$$

(2) Quantum capacitance: It originates from the quantum electrostatic energy stored in a CNT shell when it carries current. According to the Pauli exclusion principle, it is only possible to add extra electrons into the CNT shell at an available state above the Fermi level. By equating this energy to the effective capacitance energy, the quantum capacitance of each conducting channel (c_q') in a CNT can be expressed as

$$c_q' = \frac{2e^2}{hv_F} \tag{4.5c}$$

The distributed line capacitance $c_{q,ESC}$ is expressed in terms of quantum capacitance (c_q) and coupling capacitance (c_m) between shells

$$c_{equ,1} = c_{q,1} \tag{4.5d}$$

$$c_{equ,i} = \left(\frac{1}{c_{equ,i-1}} + \frac{1}{c_m^{i-1,i}}\right)^{-1} + c_{q,i}, \quad i = 2, 3, \ldots, N \tag{4.5e}$$

$$c_{q,ESC} = c_{equ,N} \tag{4.5f}$$

where

$$c_m^{i-1,i} = \frac{2\pi\varepsilon}{\ln(d_i/d_{i-1})}, \quad i = 2,3,\dots,N \tag{4.5g}$$

$$c_{q,i} = 2N_{ch,i}\frac{2e^2}{hv_F} \quad 1 \le i \le N \tag{4.5h}$$

4.3 FDTD Model of MWCNT Interconnect

The FDTD method is used to model the coupled MWCNT interconnect lines. The coupled-two interconnect lines are analyzed in this section; however, the model can be extended to coupled-N lines with a low computational cost.

4.3.1 The MWCNT Interconnect Line

The coupled-two MWCNT interconnect line structure is shown in Fig. 4.5, where r_{s1}, r_{s2} are the scattering resistances; l_{k1}, l_{k2} are the kinetic inductances; l_{e1}, l_{e2} are the magnetic inductances; c_{q1}, c_{q2} are the quantum capacitances; c_{e1}, c_{e2} are the electrostatic capacitances; and C_{L1}, C_{L2} are the load capacitances of line 1 and line 2, respectively, where all these values are mentioned in p.u.l. The parameters c_{12} and l_{12} are the p.u.l. coupling capacitances and mutual inductances, respectively

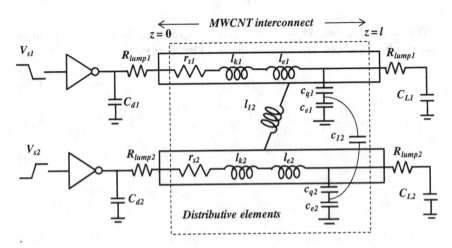

Fig. 4.5 Coupled MWCNT interconnect lines

[22–32]. The position along the interconnect line, and time are denoted as z and t, respectively.

For uniform coupled-two transmission lines the telegrapher's equations in the transverse electromagnetic (TEM) mode [11] are represented as

$$\frac{d}{dz}V(z,t) + RI(z,t) + L\frac{d}{dt}I(z,t) = 0 \tag{4.6a}$$

$$\frac{d}{dz}I(z,t) + C\frac{d}{dt}V(z,t) = 0 \tag{4.6b}$$

where V and I are 2×1 column vectors of line voltages and currents, respectively. The line parasitic elements are obtained in 2×2 per unit length matrix form, i.e.,

$$V = \begin{bmatrix} V_1 \\ V_2 \end{bmatrix}, I = \begin{bmatrix} I_1 \\ I_2 \end{bmatrix}, R = \begin{bmatrix} r_{s1} & 0 \\ 0 & r_{s2} \end{bmatrix}, L = \begin{bmatrix} l_{k1} + l_{e1} & l_{12} \\ l_{12} & l_{k2} + l_{e2} \end{bmatrix}$$ and

$$C = \begin{bmatrix} \left(1/c_{q1} + 1/c_{e1}\right)^{-1} + c_{12} & -c_{12} \\ -c_{12} & \left(1/c_{q2} + 1/c_{e2}\right)^{-1} + c_{12} \end{bmatrix}.$$

Central difference approximation is used to analyze the first-order differential Eqs. (4.6a) and (4.6b) by neglecting the higher order terms. This assumption results in a negligibly small loss of accuracy in the estimation of the transient response, since the value of time segment Δt is limited by CFL condition [33]. Using the FDTD method, the analysis of telegrapher's equations shows better accuracy, if the voltage and current points are chosen at the alternate space location and separated by one-half of the position discretization, i.e., $\Delta z/2$ [12]. In the same manner, the solution time for V and I should also be separated by $\Delta t/2$.

The interconnect line of length l is driven by a resistive driver at $z = 0$ and terminated by a capacitive load at $z = l$. The line is discretized into Nz uniform segments of length $\Delta z = l/Nz$. The voltage and current solution points are discretized along the line as shown in Fig. 4.6.

Applying finite difference approximations to (4.6a) results in

$$\frac{V_{k+1}^{n+1} - V_k^{n+1}}{\Delta z} + L\frac{I_k^{n+3/2} - I_k^{n+1/2}}{\Delta t} + R\frac{I_k^{n+3/2} + I_k^{n+1/2}}{2} = 0 \tag{4.7a}$$

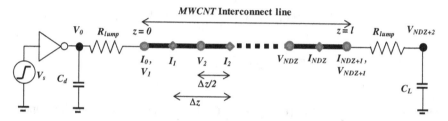

Fig. 4.6 Illustration of space discretization of line for FDTD implementation

$$I_k^{n+3/2} = EFI_k^{n+1/2} + E\left[V_k^{n+1} - V_{k+1}^{n+1}\right] \quad \text{for } k = 1, 2, \ldots, Nz \qquad (4.7b)$$

where $E = \left[\frac{\Delta z}{\Delta t}L + \frac{\Delta z}{2}R\right]^{-1}$, $F = \left[\frac{\Delta z}{\Delta t}L - \frac{\Delta z}{2}R\right]$.

Applying finite difference approximations to (4.6b) results in

$$\frac{I_k^{n+1/2} - I_{k-1}^{n+1/2}}{\Delta z} + C\frac{V_k^{n+1} - V_k^n}{\Delta t} = 0 \qquad (4.8a)$$

$$V_k^{n+1} = V_k^n + D\left[I_{k-1}^{n+1/2} - I_k^{n+1/2}\right] \quad \text{for } k = 2, 3, \ldots, Nz \qquad (4.8b)$$

where $D = \left[\frac{\Delta z}{\Delta t}C\right]^{-1}$.

4.3.2　Boundary Condition at Near-End Terminal

The voltage and current points at the near-end terminal are represented by V_1 and I_0, respectively. As indicated in Fig. 4.6, it is observed that to apply the boundary conditions in (4.8b), Δz is replaced by $\Delta z/2$. Therefore, at $k = 1$ Eq. (4.8b) becomes

$$V_1^{n+1} = V_1^n + 2D\left[I_0^{n+1/2} - I_1^{n+1/2}\right] \qquad (4.9a)$$

The source current I_0 at $(n + 1/2)$ time interval is obtained by averaging the source current at (n) and $(n + 1)$ time intervals. Then Eq. (4.9a) becomes

$$V_1^{n+1} = V_1^n + 2D\left[\frac{I_0^{n+1} + I_0^n}{2} - I_1^{n+1/2}\right] \qquad (4.9b)$$

where I_0 is the driver current. Applying Kirchhoff's current law (KCL) at near-end terminal, I_0 can be written as

$$V_0^{n+1} = V_0^n + A\left[\frac{C_m}{\Delta t}(V_s^{n+1} - V_s^n) + I_p^{n+1} - I_n^{n+1} - I_0^n\right] \qquad (4.9c)$$

$$V_1^{n+1} = BV_1^n + 2BD\left[\frac{V_0^{n+1}}{2R_{\text{lump}}} + \frac{I_0^n}{2} - I_1^{n+1/2}\right] \qquad (4.9d)$$

$$I_0^{n+1} = \frac{1}{R_{\text{lump}}}\left[V_0^{n+1} - V_1^{n+1}\right] \qquad (4.9e)$$

where $A = \left[\frac{C_m + C_d}{\Delta t}\right]^{-1}$, $B = \left[U + \frac{D}{R_{\text{lump}}}\right]^{-1}$ C_m is the drain to gate coupling capacitance, C_d is the drain diffusion capacitance of CMOS inverter, I_p and I_n are the

PMOS and NMOS currents, respectively. The modified alpha-power law model that includes the drain conductance parameter is used to express the NMOS current as

$$
I_n = \begin{cases}
0 & V_S \leq V_{\text{tn}} & \text{(off)} \\
K_{\text{ln}}(V_s - V_{\text{tn}})^{\alpha_n/2} V_0 & V_0 < V_{\text{DSATn}} & \text{(lin)} \\
K_{\text{sn}}(V_s - V_{\text{tn}})^{\alpha_n}(1 + \sigma_n V_0) & V_0 \geq V_{\text{DSATn}} & \text{(sat)}
\end{cases}
\tag{4.9f}
$$

where K_{ln}, K_{sn}, V_{tn}, α_n, and σ_n are the linear region transconductance parameter, saturation region transconductance parameter, threshold voltage, velocity saturation index, and drain conductance parameter of NMOS, respectively. In a similar manner, the PMOS current can be expressed as

$$
I_p = \begin{cases}
0 & V_S \geq V_{\text{DD}} - |V_{\text{tp}}| & \text{(off)} \\
K_{\text{lp}}(V_{\text{DD}} - V_s - |V_{\text{tp}}|)^{\alpha_p/2}(V_{\text{DD}} - V_0) & V_0 > V_{\text{DD}} - V_{\text{DSATp}} & \text{(lin)} \\
K_{\text{sp}}(V_{\text{DD}} - V_s - |V_{\text{tp}}|)^{\alpha_p}(1 + \sigma_p(V_{\text{DD}} - V_0)) & V_0 \leq V_{\text{DD}} - V_{\text{DSATp}} & \text{(sat)}
\end{cases}
\tag{4.9g}
$$

4.3.3 Boundary Condition at Far-End Terminal

Here the objective is to derive the voltage expression at $k = Nz + 1$ and $Nz + 2$. At $k = Nz + 1$, Eq. (4.8b) becomes

$$
V_{Nz+1}^{n+1} = V_{Nz+1}^n + 2D\left(I_{Nz}^{n+1/2} - \frac{I_{Nz+1}^{n+1} + I_{Nz+1}^n}{2}\right)
\tag{4.10a}
$$

Applying KCL at far-end terminal, the output current (I_{Nz+1}) can be expressed as

$$
V_{Nz+1} - V_{Nz+2} = R_{\text{lump}} I_{Nz+1}
\tag{4.10b}
$$

The discretized form of (4.10b) is

$$
I_{Nz+1}^{n+1} = \frac{1}{R_{\text{lump}}}\left[V_{Nz+1}^{n+1} - V_{Nz+2}^{n+1}\right]
\tag{4.10c}
$$

Using (4.10a) and (4.10c) the far-end voltage V_{Nz+1} can be expressed as

$$
V_{Nz+1}^{n+1} = B V_{Nz+1}^n + 2BD\left[\frac{V_{Nz+2}^{n+1}}{2R_{\text{lump}}} + I_{Nz}^{n+1/2} - \frac{I_{Nz+1}^n}{2}\right]
\tag{4.10d}
$$

and the load voltage V_{Nz+2} is

$$V_{Nz+2}^{n+1} = V_{Nz+2}^{n} + \frac{\Delta t}{C_L} I_{Nz+1}^{n} \qquad (4.10e)$$

These equations are evaluated in a bootstrapping fashion. Initially, the voltages along the line are evaluated for a specific time from Eqs. (4.9c), (4.9d), (4.8b), (4.10e), and (4.10d) in terms of the previous values of voltage and current. Thereafter, the currents are evaluated from (4.9e), (4.7b), and (4.10c) in terms of these voltages and previous current values.

4.4   Validation of the Model

The coupled MWCNT interconnects are analyzed using the actual CMOS driver. The proposed model is implemented with the MATLAB. The industry standard HSPICE simulations are used for the validation of the results. The HSPICE simulations are carried out using the subcircuit model with 50 distributed segments for interconnect and using BSIM4 technology model for MOSFET. A symmetric CMOS driver is used to drive the interconnect load. The equivalent resistance of the driver is evaluated by averaging the resistance value over an interval when the input is between V_{DD} and $V_{DD}/2$ [34]. The signal integrity analysis is carried out at the global interconnect length of 1 mm for 32 nm technology and 0.9 V of V_{DD}. The interconnect dimensions are based on the ITRS data [35]. The interconnect width and height from the ground plane are 48 and 110.4 nm, respectively. The spacing between the two interconnects is 48 nm. The relative permittivity of the inter layer dielectric medium is 2.25. The load capacitance and input transition time are 2 fF and 20 ps, respectively. The following RLC parasitics are used in the experiments [36–43]:

$$R = \begin{bmatrix} 653.67 & 0 \\ 0 & 653.67 \end{bmatrix} \frac{k\Omega}{m}, \quad L = \begin{bmatrix} 14.83 & 0.61 \\ 0.61 & 14.83 \end{bmatrix} \frac{\mu H}{m} \quad \text{and}$$

$$C = \begin{bmatrix} 93.33 & -71.50 \\ -71.50 & 93.33 \end{bmatrix} \frac{pF}{m}$$

In the interconnect system, lines 1 and 2 are considered as aggressor and victim lines, respectively. For the above-mentioned setup, the transient response is analyzed at the far-end terminal of the victim line using the proposed model, resistive driver-based model [14], and HSPICE simulations using CMOS driver. From Fig. 4.7, it can be observed that the model presented in [14] is unable to capture the timing waveform accurately. However, the proposed model is able to successfully capture the HSPICE waveform characteristics.

The crosstalk-induced delay is analyzed under two different cases. First case considers out-phase delay where the input signals of aggressor and victim lines are

Fig. 4.7 Transient response at the far-end terminal of the victim line when the aggressor and victim lines are switched out-of-phase

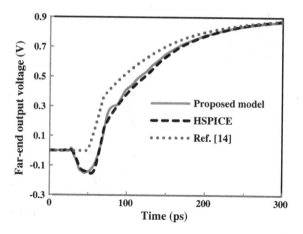

Fig. 4.8 Crosstalk-induced delay comparison **a** out-phase delay and **b** in-phase phase delay with the variation of interconnect length

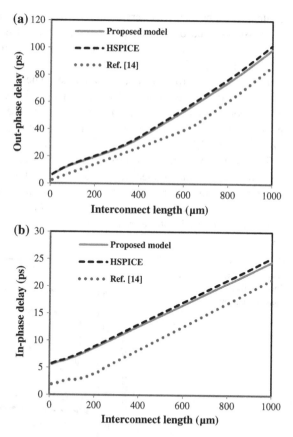

switched out-of-phase. Second case considers in-phase delay where the input signals of aggressor and victim lines are switched in-phase. Figure 4.8a, b show out-phase and in-phase delay comparison, respectively, for different interconnect

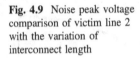

Fig. 4.9 Noise peak voltage comparison of victim line 2 with the variation of interconnect length

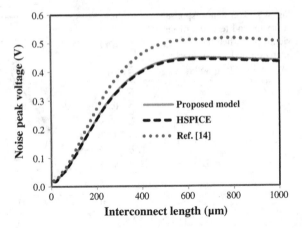

lengths. It can be clearly observed that the model proposed in [14], fails to estimate the crosstalk-induced delay for all interconnect lengths. The model proposed in [14] underestimates the delay for both out-phase and in-phase switching by average errors of 27.2 and 35.3 %, respectively.

The functional crosstalk noise is analyzed when the aggressor line is switched and the victim line is kept in quiescent mode. Figure 4.9 depicts the noise peak voltage comparison on the victim line. It can be observed that the resistive driver model [14] overestimates the noise peak voltage, wherein the average error is observed to be 15 %.

To test the robustness, the proposed model is examined at different input transition times. The interconnect length is considered as 500 μm. Figure 4.10 depicts the computational error involved in predicting the crosstalk-induced propagation delay. It can be observed that the proposed model accurately predicts the delay for both out-phase and in-phase transitions. The average error involved is only 1.4 and

Fig. 4.10 Crosstalk-induced delay comparison on victim line 2 with the variation of input transition time

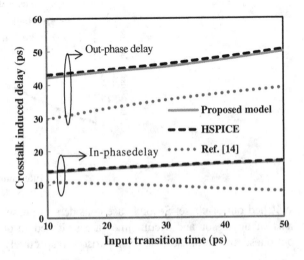

Table 4.1 CPU runtime comparison between proposed model and HSPICE with 1000 time segments

Number of space segments	HSPICE (s)	Proposed model (s)	% reduction in runtime
1	0.14	0.02	85.71
10	0.68	0.06	91.17
50	2.97	0.23	92.25
100	6.04	0.31	94.86

1.5 % during in-phase and out-phase switching, respectively. Contrastingly, with the resistive driver model [14], the average errors involved are 38.6 and 25.1 % for in-phase and out-phase switching, respectively.

Modified nodal analysis (MNA) is the core approach used in SPICE to formulate the system equations. Applying the Kirchhoff's current law and following the energy conversion principle, the MNA generates the set of matrix equations. The order of the matrix is determined by the number of nodes and unknown variables in the circuit. The unknown variables are solved after the inversion of the matrix and therefore require more computational time. However, the FDTD operator is matrix free and therefore fast and memory efficient as compared to HSPICE simulations.

The efficiency of the proposed model is examined under different test cases. The analysis is carried out by varying the space segment while keeping the time segment constant for coupled interconnects. Using a PC with Intel Dual Core CPU (2.33 GHz, 4 GB RAM), the comparison results are provided in Table 4.1. Using the proposed model, it is observed that the CPU runtime reduces by an average of 91 % in comparison to HSPICE simulations. Additionally, the proposed model is compared with the HSPICE simulations using the same modified alpha-power law model. It is observed that the average CPU runtime reduces by 88 % in comparison to HSPICE simulations.

4.5 Sensitivity Analysis

The primary assumptions made in the proposed work are for: (1) number of conducting channels and (2) contact resistance. This subsection presents the sensitivity analysis to evaluate the validity of these assumptions.

4.5.1 Sensitivity Analysis for Number of Conducting Channels

The number of conducting channels in a CNT can be obtained from expression (4.2b), which is an approximated form of (4.2a). Table 4.2 shows the variations in

Table 4.2 Variation between (4.2a) and (4.2b) on parasitics and crosstalk-induced performance parameters

Variation between (4.2a) and (4.2b)	Parasitic parameter				Performance parameter		
	Lumped resistance (Ω)	Scattering resistance (Ω)	C (fF)	L (pH)	Noise peak voltage (V)	In-phase delay (ps)	Out-phase delay (ps)
From Eq. (4.2a)	11.79	675.10	21.84	15.28	0.433	26.0	64.3
From Eq. (4.2b)	11.43	653.67	21.83	14.83	0.424	25.3	63.6
% change	3.05	3.17	0.04	2.94	2.1	2.6	1.1

parasitics and crosstalk-induced performance parameters using Eqs. (4.2a) and (4.2b). The average percentage change in parasitics and performance parameters are just 2.3 and 2 %, respectively. It can be inferred that the parasitics and crosstalk-induced performance parameters are almost insensitive to the usage of approximated expression for obtaining N_{ch} [44].

4.5.2   Sensitivity Analysis for Contact Resistance

The value of imperfect metal contact resistance can range from the best case value of zero to the worst case value of few kilo-ohms depending on the fabrication process. As reported earlier [9], the R_{MC} value is considered as 3.2 kΩ per shell. However, a sensitivity analysis on parasitic $R_{lump,ESC}$ and crosstalk-induced performance parameters for R_{MC} varying from 0 to 8 kΩ is carried out and the results are presented in Table 4.3. A maximum variation of 5 % in $R_{lump,ESC}$ and almost no change in the crosstalk performance are noticed with the change in R_{MC}. This is due to the fact that the crosstalk-induced performance parameters primarily depend on the scattering resistance and almost insensitive to the change in R_{MC}.

Table 4.3 Variation of performance parameters due to change in R_{MC}

Parasitic parameter		Performance parameter			
R_{MC} (per shell) (k Ω)	Lumped resistance ($R_{lump,ESC}$) (Ω)	Noise peak voltage (V)	Noise peak timing (ps)	In-phase delay (ps)	Out-phase delay (ps)
0	11.20	0.425	54.5	25.1	63.5
2	11.3	0.425	54.4	25.3	63.5
4	11.49	0.424	54.4	25.3	63.6
6	11.63	0.424	54.4	25.4	63.6
8	11.77	0.424	54.4	25.4	63.8

4.6 Summary

This chapter presented an accurate model to analyze the crosstalk effects in coupled MWCNT interconnect lines. The CMOS driver and the coupled MWCNT interconnect are modeled by modified alpha-power law model and FDTD method, respectively. It has been observed that the results of the proposed model exhibit a good agreement with HSPICE simulations. Over the random number of test cases, the average error in the propagation delay measurement is observed to be less than 2 %. Moreover, the sensitivity analysis is performed based on the assumptions used in the proposed model. It is observed that the percentage change in parasitic elements and performance parameters are almost negligible with respect to the assumptions associated with the model. This analysis suggests that with continuous advancements in FDTD technique the proposed model would play a significant role in performance analysis of MWCNT on-chip interconnects and would be potentially incorporated in TCAD simulators.

References

1. D'Amore M, Sarto MS, Tamburrano A (2010) Fast transient analysis of next-generation interconnects based on carbon nanotubes. IEEE Trans Electromagn Compat 52(2):496–503
2. Li H, Xu C, Srivastava N, Banerjee K (2009) Carbon nanomaterials for next-generation interconnects and passives: physics, status and prospects. IEEE Trans Electron Devices 56 (9):1799–1821
3. Sahoo M, Rahaman H (2013) Modeling of crosstalk delay and noise in single-walled carbon nanotube bundle interconnects. In: Proceedings of annual IEEE india conference (INDICON 2013), Mumbai, India, pp 1–6
4. Das D, Rahaman H (2010) Timing analysis in carbon nanotube interconnects with process, temperature, and voltage variations. In: Proceedings of IEEE international symposium electronic design (ISED 2010), Bhubaneshwar, India, pp 27–32
5. McEuen PL, Fuhrer MS, Park H (2002) Single-walled carbon nanotube electronics. IEEE Trans Nanotechnol 1(1):78–85
6. Naeemi A, Meindl JD (2009) Carbon nanotube interconnects. Annu Rev Mater Res 39 (1):255–275
7. Li HJ, Lu WG, Li JJ, Bai XD, Gu CZ (2005) Multichannel ballistic transport in multiwall carbon nanotubes. Phys Rev Lett 95(8):86601
8. Sarto MS, Tamburrano A (2010) Single conductor transmission-line model of multiwall carbon nanotubes. IEEE Trans Nanotechnol 9(1):82–92
9. Li H, Yin WY, Banerjee K, Mao JF (2008) Circuit modeling and performance analysis of multi-walled carbon nanotube interconnects. IEEE Trans Electron Devices 55(6):1328–1337
10. Tang M, Lu J, Mao J (2012) Study on equivalent single conductor model of multi-walled carbon nanotube interconnects. In: proceedings of IEEE Asia pacific microwave conference, Taiwan, pp 1247–1249
11. Paul CR (2008) Analysis of multiconductor transmission lines. IEEE Press
12. Paul CR (1994) Incorporation of terminal constraints in the FDTD analysis of transmission lines. IEEE Trans Electromagn Compat 36(2):85–91
13. Paul CR (1996) Decoupling the multi conductor transmission line equations. IEEE Trans Microw Theory Tech 44(8):1429–1440

14. Liang F, Wang G, Lin H (2012) Modeling of crosstalk effects in multiwall carbon nanotube interconnects. IEEE Trans Electromagn Compat 54(1):133–139
15. Xu T, Wang Z, Miao J, Chen X, Tan CM (2007) Aligned carbon nanotubes for through-wafer interconnects. Appl Phys Letts 91(4):042108-1–042108-3
16. Wang Z, Chen X, Zhang J, Tang N, Cai J (2013) Fabrication of sensor based on MWCNT for NO_2 and NH_3 detection. In: Proceedings of IEEE Conference on Nanotechnology, Beijing, pp 2202–2214
17. Srivastava A, Xu Y, Sharma AK (2010) Carbon nanotubes for next generation very large scale integration interconnects. J Nanophotonics 4(1):1–26
18. Xu Y, Srivastava A (2009) A model for carbon nanotube interconnects. Int J Circuit Theory Appl 38(6):559–575
19. Somvanshi D, Jit S (2014) Effect of ZnO Seed layer on the electrical characteristics of Pd/ZnO thin film based Schottky contacts grown on n-Si substrates. IEEE Trans Nanotechnol 13 (6):1138–1144
20. Burke PJ (2002) Lüttinger liquid theory as a model of the gigahertz electrical properties of carbon nanotubes. IEEE Trans Nanotechnol 1(3):129–144
21. Park JY, Rosenblatt S, Yaish Y, Sazonova V, Üstünel H, Braig S, Arias TA, Brouwer PW, McEuen PL (2004) Electron—phonon scattering in metallic single-walled carbon nanotubes. Nano Lett 4(3):517–520
22. Harris PJF (1999) Carbon nanotubes and relayed structures: new materials for 21st century. Press syndicate of the university of cambridge, Cambridge, United Kingdom
23. Collins PG, Hersam M, Arnold M, Martel R, Avouris Ph (2001) Current saturation and electrical breakdown in multiwalled carbon nanotubes. Phys Rev Lett 86(14):3128–3131
24. Li H, Banerjee K (2009) High-frequency analysis of carbon nanotube interconnects and implications for on-chip inductor design. IEEE Trans Electron Devices 56(10):2202–2214
25. Naeemi A, Meindl JD (2007) Physical modeling of temperature coefficient of resistance for single- and multi-wall carbon nanotube interconnects. IEEE Electron Device Lett 28 (2):135–138
26. Naeemi A, Meindl JD (2008) Performance modeling for single- and multiwall carbon nanotubes as signal and power interconnects in gigascale systems. IEEE Trans Electron Devices 55(10):2574–2582
27. Maffucci A, Miano G, Villone F (2009) A new circuit model for carbon nanotube interconnects with diameter-dependent parameters. IEEE Trans Nanotechnol 8(3):345–354
28. Nieuwoudt A, Massoud Y (2006) Evaluating the impact of resistance in carbon nanotube bundles for VLSI interconnect using diameter-dependent modeling techniques. IEEE Trans Electron Devices 53(10):2460–2466
29. Kim W, Javey A, Tu R, Cao J, Wang Q, Dai H (2005) Electrical contacts to carbon nanotubes down to 1 nm in diameter. Appl Phys Lett 87(17):173101
30. Nieuwoudt A, Massoud Y (2006) Understanding the impact of inductance in carbon nanotube bundles for VLSI interconnect using scalable modeling techniques. IEEE Trans Nanotechnol 5 (6):758–765
31. Nieuwoudt A, Massoud Y (2007) Performance implications of inductive effects for carbon-nanotube bundle interconnect. IEEE Electron Device Lett 28(4):305–307
32. Raychowdhury A, Roy K (2006) Modeling of metallic carbon-nanotube interconnects for circuit simulations and a comparison with Cu interconnects for scaled technologies. IEEE Trans Comput Aided Des Integr Circ Syst 25(1):58–65
33. Courant R, Friedrichs K, Lewy H (1967) On the partial difference equations of mathematical physics. IBM J Res Devel 11(2):215–234
34. Rabaey JM, Chandrakasan A, Nikolic B (2003) Digital integrated circuits: a design perspective, 2nd edn. Prentice-Hall, New Jersey
35. International Technology Roadmap for Semiconductors (2013) http://public.itrs.net (Online)
36. Agrawal S, Raghuveer MS, Ramprasad R, Ramanath G (2007) Multishell carrier transport in multiwalled carbon nanotubes. IEEE Trans Nanotechnol 6(6):722–726

37. Nieuwoudt A, Massoud Y (2008) On the optimal design, performance, and reliability of future carbon nanotube-based interconnect solutions. IEEE Trans Electron Devices 55(8):2097–2110
38. Maffucci A, Miano G, Villone F (2008) Performance comparison between metallic carbon nanotube and copper nano-interconnects. IEEE Trans Adv Packag 31(4):692–699
39. Fathi D, Forouzandeh B, Mohajerzadeh S, Sarvari R (2009) Accurate analysis of carbon nanotube interconnects using transmission line model. Micro Nano Lett 4(2):116–121
40. Lim SC, Jang JH, Bae DJ, Han GH, Lee S, Yeo IS, Lee YH (2009) Contact resistance between metal and carbon nanotube interconnects: effect of work function and wettability. Appl Phys Lett 95(26):264103-1–264103-3
41. Koo KH, Cho H, Kapur P, Saraswat KC (2007) Performance comparison between carbon nanotubes, optical and Cu for future high-performance on-chip interconnect applications. IEEE Trans Electron Devices 54(12):3206–3215
42. Bellucci S, Onorato P (2010) The role of the geometry in multiwall carbon nanotube interconnects. J Appl Phys 108(7):073704-1–073704-9
43. Rossi D, Cazeaux JM, Metra C, Lombardi F (2007) Modeling crosstalk effects in CNT bus architecture. IEEE Trans Nanotechnol 6(2):133–145
44. Kumar VR, Kaushik BK, Patnaik A (2015) Crosstalk noise modeling of multiwall carbon nanotube (MWCNT) interconnects using finite-difference time-domain (FDTD) technique. Microelectron Reliab 55(1):155–163

Chapter 5
Crosstalk Modeling with Width Dependent MFP in MLGNR Interconnects Using FDTD Technique

Abstract This chapter analyzes the performance of coupled MLGNR interconnects using the FDTD technique. In a more realistic manner, the proposed model incorporates the width dependent MFP parameter of the MLGNR while taking into account the edge roughness. This helps in accurate estimation of the crosstalk-induced performance in comparison to the conventional models. The crosstalk noise is comprehensively analyzed by examining both functional and dynamic crosstalk effects.

Keywords Equivalent single conductor (ESC) model · In-phase and out-phase delay · Mean free path (MFP) · Multilayer graphene nanoribbon (MLGNR) · Propagation delay · Power dissipation

5.1 Introduction

In the first four decades of the semiconductor industry, system performance was entirely dependent on the transistor delay and power dissipation [1]. The technology scaling had an adverse effect on *RLC* delay of complex VLSI circuits as the resistivity increased for the small-dimensional metal interconnects made up of Cu [2]. The reduced cross-sectional area of Cu interconnects resulted in higher resistivity under the effects of enhanced grain and surface scattering. Moreover, with thinner interconnects and higher operating frequency, electromigration-induced problems gained more attention. Presently, at GHz range of frequencies, issues like skin effect, stability, operational bandwidth and crosstalk severely affect the performance of Cu interconnects [3]. Therefore, researchers are forced to find an alternative to Cu material for high-speed global VLSI interconnects [4, 5].

During the recent past, graphene nanoribbons (GNRs) have rapidly gained importance as an emerging material that potentially forms a monolithic system for field effect devices and interconnects [6, 7]. GNR is a sheet of graphite wherein carbon atoms are tightly packed in honeycomb lattice structures [8]. High-quality GNR sheets have long mean free path (MFP) ranging from 1 to 5 µm that results in

© The Author(s) 2016
B.K. Kaushik et al., *Crosstalk in Modern On-Chip Interconnects*,
SpringerBriefs in Applied Sciences and Technology,
DOI 10.1007/978-981-10-0800-9_5

ballistic transport phenomenon. Due to the large MFP, GNRs have higher carrier mobility of 10^5 cm^2/(V·s) and larger current densities, of 10^9 A/cm^2 in comparison to Cu [9]. Due to high intrinsic resistance of single layer GNR, researchers often prefer multilayer GNR (MLGNR) as potential interconnect material [10–12]. Moreover, intercalation doping can increase the in-plane conductivity of MLGNR up to 20 times that involves insertion of one dopant layer between each pair of adjacent graphene layers [8]. Intercalation doping can also increase the MFP due to an increase in spacing between the adjacent layers. Additionally, the easier fabrication process of MLGNR makes it a promising candidate for VLSI interconnect material. The comparison between the performance of MLGNR and Cu has been studied in [13], where the authors observed that MLGNR interconnect demonstrates the smaller propagation delay than Cu interconnect.

Using the equivalent transmission line model, the crosstalk effects of coupled MLGNR have been studied in [14], where the authors considered the MFP parameter independent of width by assuming perfectly smooth edges of MLGNR. However, in reality all GNRs exhibit edge roughness [15, 16]. Due to these rough edges, the electron scattering increases, thereby decreasing the overall MFP and increasing the resistivity [17]. At lower widths, the MFP is predominantly dependent on the edge roughness. Therefore, it is essential to incorporate width dependent MFP while modeling the performance of MLGNR-based interconnects.

This chapter accurately analyzes the performance of MLGNR interconnects based on the finite-difference time-domain (FDTD) technique. In a more realistic manner, the proposed model includes the effect of width dependent MFP of the MLGNR while taking into account the edge roughness. Moreover, a nonlinear CMOS driver is used to drive the MLGNR interconnect line. At different interconnect widths, the crosstalk-induced propagation delay is compared among proposed model, HSPICE, and the existing crosstalk noise model.

The organization of this chapter is as follows: Sect. 5.1 introduces the importance of MLGNR interconnects in current research scenario and briefs about the work carried out. Based on the multiconductor transmission line theory, an equivalent single conductor (ESC) model of MLGNR interconnects is described in Sect. 5.2. Using a driver-interconnect-load system, a comparative analysis of transient response of MTL and ESC models is also presented in this section. Section 5.3 brief the FDTD model for the MLGNR interconnects. The validation of the proposed model is discussed in Sect. 5.4 along with the performance comparison between Cu and MLGNR interconnects. Section 5.5 concludes this chapter.

5.2 Equivalent Single Conductor Model of the MLGNR Interconnect

The proposed model is developed for MLGNR interconnect line positioned over a ground plane at a distance H with a dielectric medium sandwiched between GNRs as shown in Fig. 5.1. The MLGNR consists of N number of layers

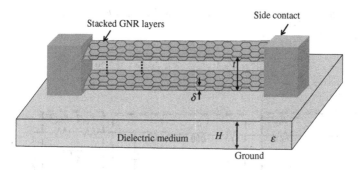

Fig. 5.1 The geometric structure of MLGNR

$$N_{\text{layer}} = 1 + \text{int}[t/\delta] \tag{5.1}$$

where w, t and δ are width, thickness, and interlayer spacing, respectively. The number of conducting channels per layer can be expressed as

$$N_{\text{ch}} = \sum_{i=0}^{n_c} \left[1 + e^{(E_i - E_F)/k_B T}\right]^{-1} + \sum_{i=0}^{n_v} \left[1 + e^{(E_i + E_F)/k_B T}\right]^{-1} \tag{5.2a}$$

where the first and second summations represent the contributions of the n_c conduction subbands and n_v valence subbands, respectively, T is the temperature, k_B is the Boltzmann constant, E_F is the Fermi level, and E_i is the lowest (highest) energy of the ith conduction (valence) subband. In general, the value of E_F is set to zero for neutral GNR [8]. However, some charge usually gets trapped at the interface of graphene and the substrate. This is due to the planar structure of graphene and also due to the work function difference between the graphene and substrate. Alternatively, the number of conduction channels is also derived from the following approximated expression [18]

$$N_{\text{ch}} = \begin{cases} a_0 + a_1 w + a_2 w^2 + a_3 E_F + a_4 w E_F + a_5 E_F^2 & \text{for } E_F > 0 \\ b_0 + b_1 + b_2 w^2 & \text{for } E_F = 0 \end{cases} \tag{5.2b}$$

For metallic GNR at $T = 300$ K, the fitting parameters a_0–a_6 and b_0–b_2 are given in Tables 5.1 and 5.2, respectively [18].

The typical ESC model of an MLGNR is shown in Fig. 5.2, where R_{MC}, R_Q, and r_s are the imperfect metal contact, quantum, and scattering resistances, respectively; l_k

Table 5.1 Fitting parameters (a_0–a_6) for calculating the N_{ch} of $E_F > 0$

Fitting parameters					
a_0	a_1	a_2	a_3	a_4	a_5
1.244	-1.696×10^{-2}	7.517×10^{-5}	-5.031	1.225	5.122

Table 5.2 Fitting parameters (b_0–b_3) for calculating the N_{ch} of $E_F = 0$

Fitting parameters		
b_0	b_1	b_2
1.94	2.97×10^{-4}	2.29×10^{-4}

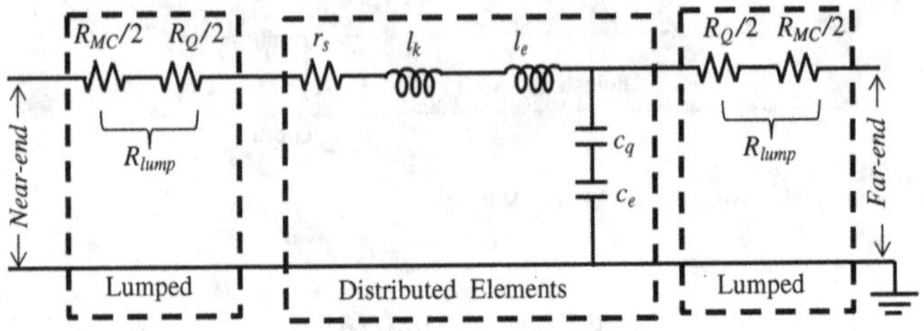

Fig. 5.2 Equivalent single conductor model of an MLGNR interconnect

and l_e are the kinetic and magnetic inductances, respectively; c_q and c_e are the quantum and electrostatic capacitances, respectively. The R_{lump} represents the average value of metal contact resistance and quantum resistance.

The resistances R_{lump} and r_s are expressed as

$$R_{\text{lump}} = \frac{1}{2}\left[\frac{h/2e^2}{N_{\text{ch}}N_{\text{layer}}} + \frac{R_{\text{MC}}}{N_{\text{layer}}}\right] \tag{5.3}$$

$$r_{s,\text{ESC}} = \left(h/2e^2 N_{\text{layer}}\right)\left(\sum_n \left(\frac{l}{\lambda_{\text{eff}_n}}\right)^{-1}\right)^{-1} \tag{5.4}$$

where h, e, N_{layer}, n, l and λ_{eff} represent the Planck's constant, electron charge, number of GNR layers, number of subbands, length, and overall effective MFP, respectively. The λ_{eff} of nth subband is expressed as

$$\frac{1}{\lambda_{\text{eff}_n}} = \frac{1}{\lambda_d} + \frac{1}{\lambda_n} \tag{5.5}$$

where λ_d and λ_n represent the MFP corresponding to the scattering effects due to defects and edge roughness, respectively. The value of λ_d is considered as 419 nm and 1.03 μm for neutral and doped MLGNRs, respectively. The λ_n for nth subband is expressed as [17]

$$\lambda_n = \frac{w}{P}\sqrt{\left(\frac{2wE_F}{nhv_F}\right)^2 - 1} \tag{5.6}$$

where P is the backscattering probability, lies in the range 0–1 and v_F is the Fermi velocity.

The quantum capacitance and kinetic inductance of an MLGNR can be expressed as

$$c_{q,\text{ESC}} = N_{\text{ch}} N_{\text{layer}} \frac{2 \times 2q^2}{h v_F} \tag{5.7}$$

$$l_{k,\text{ESC}} = \frac{1}{N_{\text{ch}} N_{\text{layer}}} \frac{h}{2 \times 2q^2 v_F} \tag{5.8}$$

The values of $l_{e,\text{ESC}}$ and $c_{e,\text{ESC}}$ can be obtained using the electromagnetic field solvers.

5.2.1 Transient Analysis of MTL and ESC Models

The ESC model of MLGNR is validated with respect to the MTL model by performing transient analysis of DIL system. The CMOS gate-driven MLGNR interconnect line is shown in Fig. 5.3. The number of stacked GNR layers is considered as 20. The interconnect line is excited and terminated by a CMOS driver and capacitive load, respectively. The symmetric CMOS inverter is used and the load capacitance is considered as 250 aF. To maintain good accuracy, the number of distributed segments is considered as 20. For different interconnect lengths ranging from 100 to 1000 μm, Fig. 5.4 shows the far-end voltage waveforms of MLGNR interconnects. It is observed that the output voltage waveforms of the ESC model are in good agreement with the MTL model for all interconnect lengths.

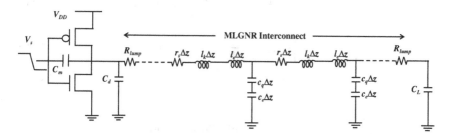

Fig. 5.3 Driver-interconnect-load (DIL) structure, wherein R_{lump} is placed at near-end and far-end terminals of the interconnect line due to the effect of quantum and imperfect contact resistances

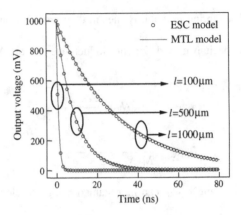

Fig. 5.4 Transient waveforms of the output voltages of MLGNR (N_{layer} = 20) interconnect

5.3   FDTD Model of the MLGNR Interconnect

The MLGNR interconnect of length l is driven by a CMOS inverter at near end and terminated by a capacitive load at far end. The total interconnect length is discretized into Nz uniform segments of space step Δz and the total simulation time is discretized into n uniform segments of time step Δt. The value of n can be determined by dividing the total simulation time by Δt. The time step, Δt is determined by the Courant stability condition. The maximum time step that can be allowed for the stable operation is $\Delta t_{max} = \Delta z / v_{max}$, where v_{max} is the maximum phase velocity. The voltage and current solution points are discretized along the line as shown in Fig. 5.5.

The CMOS gate-driven coupled MWCNT interconnects are shown in Fig. 5.6. The telegrapher's equations are

$$\frac{\mathrm{d}}{\mathrm{d}z}\boldsymbol{V}(z,\,t) + \boldsymbol{R}\boldsymbol{I}(z,\,t) + \boldsymbol{L}\frac{\mathrm{d}}{\mathrm{d}t}\boldsymbol{I}(z,\,t) = 0 \qquad (5.9a)$$

$$\frac{\mathrm{d}}{\mathrm{d}z}\boldsymbol{I}(z,\,t) + \frac{\mathrm{d}}{\mathrm{d}t}\boldsymbol{C}\boldsymbol{V}(z,\,t) = 0 \qquad (5.9b)$$

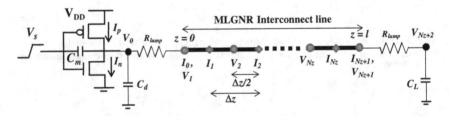

Fig. 5.5 Illustration of space discretization of line for FDTD implementation

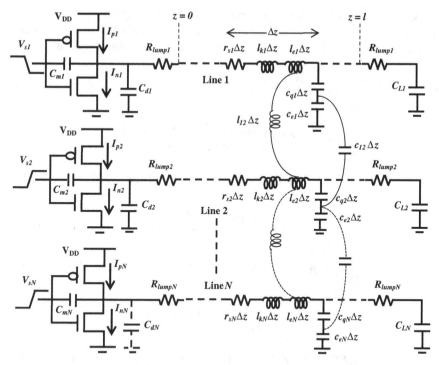

Fig. 5.6 Coupled MWCNT interconnect lines driven by CMOS inverter

where V and I are $N \times 1$ column vectors and the line parasitic elements are obtained in $N \times N$ matrix form. For instance, the voltage V is $[\, V_1 \quad V_2 \quad . \quad . \quad . \quad V_{N-1} \quad V_N \,]^T$,

resistance R, and inductance L matrices are $\begin{bmatrix} r_{s1} & 0 & 0 & . & . \\ 0 & r_{s2} & 0 & . & . \\ 0 & 0 & r_{s3} & . & . \\ . & . & . & . & . \\ . & . & . & 0 & r_{sN} \end{bmatrix}$ and

$\begin{bmatrix} l_{k1}+l_{e1} & l_{12} & l_{13} & . & . \\ l_{21} & l_{k2}+l_{e2} & l_{23} & . & . \\ l_{31} & l_{32} & l_{k3}+l_{e3} & . & . \\ . & . & . & . & . \\ . & . & . & l_{N-1,N} & l_{kN}+l_{eN} \end{bmatrix}$, respectively, where r_{sN} is the

scattering resistance of line N, and $l_{N-1,N}$ is the mutual inductance between the lines $N-1$ and N.

Applying finite difference approximations to (5.9a) and (5.9b) the line voltages and currents can be determined, whereas using Kirchhoff's law at near-end and far-end boundaries the terminal voltage and current can be determined as

$$V_0^{n+1} = V_0^n + A\left[\frac{C_m}{\Delta t}\left(V_s^{n+1} - V_s^n\right) + I_p^{n+1} - I_n^{n+1} - I_0^n\right] \tag{5.10}$$

$$V_1^{n+1} = BV_1^n + 2BD\left[\frac{V_0^{n+1}}{2R_{lump}} + \frac{I_0^n}{2} - I_1^{n+1/2}\right] \tag{5.11}$$

$$V_k^{n+1} = V_k^n + D\left[I_{k-1}^{n+1/2} - I_k^{n+1/2}\right] \text{ for } k = 2, 3, \ldots, Nz \tag{5.12}$$

$$V_{Nz+2}^{n+1} = V_{Nz+2}^n + \frac{\Delta t}{C_L}I_{Nz+1}^n \tag{5.13}$$

$$V_{Nz+1}^{n+1} = BV_{Nz+1}^n + 2BD\left[\frac{V_{Nz+2}^{n+1}}{2R_{lump}} + I_{Nz}^{n+1/2} - \frac{I_{Nz+1}^n}{2}\right] \tag{5.14}$$

$$I_0^{n+1} = \frac{1}{R_{lump}}\left[V_0^{n+1} - V_1^{n+1}\right] \tag{5.15}$$

$$I_k^{n+3/2} = EFI_k^{n+1/2} + E\left[V_k^{n+1} - V_{k+1}^{n+1}\right] \quad \text{for } k = 1, 2, \ldots\ldots, Nz \tag{5.16}$$

$$I_{Nz+1}^{n+1} = \frac{1}{R_{lump}}\left[V_{Nz+1}^{n+1} - V_{Nz+2}^{n+1}\right] \tag{5.17}$$

where $A = \left[\frac{C_m+C_d}{\Delta t}\right]^{-1}$, $B = \left[U + \frac{D}{R_{lump}}\right]^{-1}$, $D = \left[\frac{\Delta z}{\Delta t}C\right]^{-1}$, $E = \left[\frac{\Delta z}{\Delta t}L + \frac{\Delta z}{2}R\right]^{-1}$, $F = \left[\frac{\Delta z}{\Delta t}L - \frac{\Delta z}{2}R\right]$, C_m is the drain to gate coupling capacitance, C_d is the drain diffusion capacitance of CMOS inverter, I_p and I_n are the PMOS and NMOS currents, respectively [19].

The expressions (5.10)–(5.17) are evaluated in a bootstrapping fashion. The process can be demonstrated with the flowchart shown in Fig. 5.7. The proposed model can be easily extended to bigger circuits with more number of coupled lines by changing the dimensions of the parasitic, voltage, and current matrices. To optimize the calculation time in the FDTD flow, the value of Δt must be considered as Δt_{max}, so that the number of time steps can be minimized.

Fig. 5.7 Flowchart for
evaluation of voltage and
current using the proposed
FDTD technique

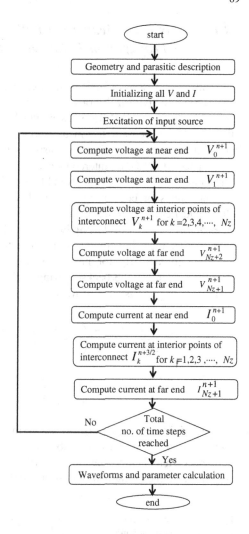

5.4 Results and Discussion

The effective mean free path, resistance, propagation delay, and the
crosstalk-induced delay of an MLGNR are analyzed by incorporating the width
dependent MFP. Moreover, the performance of MLGNR interconnect is compared
with Cu interconnect.

5.4.1 Analysis of Mean Free Path, Resistance, and Propagation Delay with Rough Edges

The fundamental issue in GNR is the presence of edge roughness that substantially reduces the effective MFP. For rough edges, the electrons scatter at the edges, and hence the effective MFP becomes width dependent. Therefore, width dependent MFP should be incorporated in the model of MLGNR resistance. Based on the improved model, this section demonstrates the impact of edge roughness on the performance of MLGNR interconnects.

The effective MFP, λ_{eff} is analyzed by considering the width dependent MFP for different values of edge roughness probabilities of MLGNR. The MFP due to defects and Fermi level are assumed to be 419 nm and 0.2 eV, respectively. The values of λ_{eff} are calculated using Eq. (5.5) and the results are shown in Fig. 5.8. It is observed that the edge roughness reduces the MFP by more than one order of magnitude, particularly for narrow widths. However, the reduction of MFP is highly dependent on the backscattering probability (P).

The scattering resistance of MLGNR can be expressed as

$$r_{s,\text{ESC}} = \frac{h}{2e^2 N_{\text{layer}}} \left[\sum_n \left(\frac{l}{\lambda_{\text{eff}_n}} \right)^{-1} \right]^{-1} ; 0 < P \leq 1 \text{ (rough edges)} \qquad (5.18a)$$

$$r_{s,\text{ESC}} = \frac{h}{2e^2 N_{\text{layer}} N_{\text{ch}}} \left(\frac{l}{\lambda_d} \right); P = 0 \text{ (smooth edges)} \qquad (5.18b)$$

Using (5.18a) and (5.18b), the scattering resistance of MLGNR for different widths and edge roughness probabilities is shown in Fig. 5.9. The thickness and interconnect length are 56.9 nm and 100 μm, respectively. The variation in the resistance is higher at narrow widths due to the dominating effect of edge scattering.

Fig. 5.8 MFP of MLGNR for first two lowest subbands at different widths

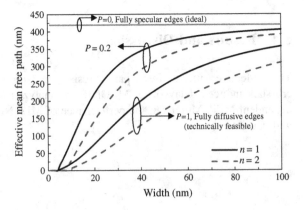

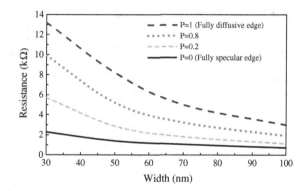

Fig. 5.9 Resistance of MLGNR for different interconnect widths

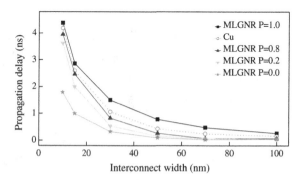

Fig. 5.10 Propagation delay of MLGNR and Cu at $t = 56.9$ nm and $l = 100$ μm

For different values of P, the propagation delay of MLGNR and Cu at different interconnect widths is shown in Fig. 5.10. It is observed that for wider MLGNR interconnects, the delay is almost constant for different edge roughness probabilities. It is due to a small variation in the scattering resistance for wider interconnects. It can also be observed that the propagation delay of MLGNR is higher than Cu for fully diffusive edge ($P = 1$).

5.4.2 Crosstalk-Induced Delay

The coupled MLGNR interconnects performance is analyzed using the proposed model and the outcome is compared with the previously reported model [14], where the MFP is considered as independent of width. An industry standard HSPICE simulator is used for model validation. Considering the width dependent and width independent MFPs, the variation of crosstalk-induced propagation delay with interconnect width during in-phase and out-phase transitions is shown in Fig. 5.11a, b, respectively. The dynamic crosstalk is analyzed by switching both lines in-phase and out-phase. The input signal rise and fall transition times are considered as 20 ps. For an interconnect length of 10 μm, the interconnect width is varied from 10 to 60 nm, while

Fig. 5.11 Crosstalk-induced propagation delay performance with change in interconnect width under **a** in-phase switching, and **b** out-phase switching

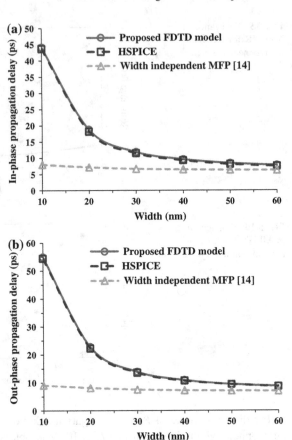

keeping the thickness and spacing between the interconnect lines as 22 nm, and distance from the ground plane as 44 nm.

From Fig. 5.11, it can be observed that the width independent MFP model presented in [14] underestimates the propagation delay by 32 %. Moreover, it is observed that this margin increases substantially for technology nodes having narrower interconnect widths. This is due to the prominent effect of edge scattering at smaller dimensions. Therefore, it is strongly recommended to include width dependent MFP for accurately modeling the crosstalk noise in MLGNR interconnects.

To further verify the robustness of the proposed model, the propagation delay comparison under in-phase and out-phase transitions on victim line 2 is observed in Fig. 5.12 for different input transition times. It can be observed that the proposed model matches accurately with HSPICE simulations for all transition times. Moreover, the propagation delay during the out-phase transition is higher due to Miller capacitance effect.

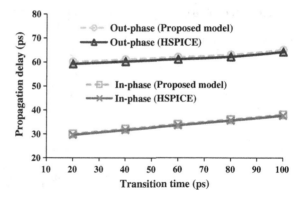

Fig. 5.12 Variation of propagation delay with respect to input transition time

Using a PC with Pentium Dual Core CPU (2.33 GHz, 4 GB RAM), the runtime of the proposed model is compared with the HSPICE simulation time. Using the proposed model, it is observed that the CPU runtime reduces by an average of 95 % in comparison to HSPICE simulations. For an interconnect length of 200 μm with 100-space and 1000-time segments, the runtime using proposed model is 0.36 s against 7.26 s using HSPICE.

5.4.3 Performance Comparison Between Cu and MLGNR Interconnects

The propagation delay and power dissipation of MLGNR (neutral and doped) and Cu are analyzed for similar interconnect length, width, and thickness. The mean free path, λ_d of neutral MLGNR is considered as 419 nm with in-plane conductivity of 0.026 (μΩ cm)$^{-1}$, whereas an AsF$_5$ doped MLGNR can exhibit λ_d of 1.03 μm with in-plane conductivity of 0.63 (μΩ cm)$^{-1}$ [8]. Figure 5.3 shows the DIL system wherein the interconnect line is represented by the ESC model of MLGNR. The *RLC* distributed model is used for analyzing the conventional Cu interconnects [20–22]. A CMOS driver with supply voltage of 0.9 V is used to drive the interconnect line that is terminated by a load capacitance of 10 fF. Using this setup, propagation delay and power dissipation are analyzed for different global interconnect lengths ranging from 100 to 1000 μm.

The propagation delay and power dissipation are proportional to the resistive and capacitive parasitics of interconnect. For different interconnect thicknesses, the MLGNR (neutral and doped) to Cu delay and power dissipation ratio are shown in Fig. 5.13a, b, respectively. It is observed that a thicker doped MLGNR demonstrates substantial reduction in delay and power dissipation compared to Cu interconnects. In doped MLGNR, the higher carrier concentration in each layer

Fig. 5.13 MLGNR to Cu
a delay and **b** power
dissipation ratio at different
interconnect thicknesses

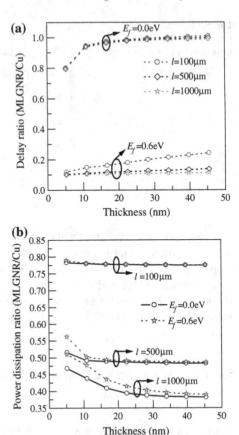

substantially increases the number of conducting channels that in turn drastically reduces the resistive parasitic ($r_{s,ESC}$) compared to Cu interconnects. Although, more number of conducting channels in doped MLGNR increases the quantum capacitance ($c_{q,ESC}$), but the equivalent capacitance (c_{ESC}) remains almost constant due to the dominating effect of ($c_{e,ESC}$) factor. Therefore, the cumulative effect of $r_{s,ESC}$ and c_{ESC} of doped MLGNR reduces the overall delay and power dissipation in comparison to Cu interconnects.

Table 5.3 summarizes the percentage reduction in propagation delay and power dissipation of doped MLGNR in comparison to Cu at different interconnect lengths. The reduction is more pronounced for longer interconnects due to a comparatively higher reduction in the line resistance. The overall delay and power dissipation of doped MLGNR are reduced by 86.13 and 43.72 %, respectively, in comparison to the Cu interconnects.

Table 5.3 Percentage reduction in propagation delay and power dissipation for doped MLGNR with respect to Cu interconnects

Thickness (nm)	% reduction in propagation delay of MLGNR *w.r.t.* Cu for interconnect lengths of			% reduction in power dissipation of MLGNR *w.r.t.* Cu for interconnect lengths		
	100 μm	500 μm	1000 μm	100 μm	500 μm	1000 μm
5.75	88.01	89.79	90.11	49.39	50.61	54.04
11.50	85.13	88.97	89.61	50.93	52.98	57.17
17.25	83.78	88.44	89.42	51.12	54.17	59.14
23.00	81.91	87.98	89.27	51.32	55.82	60.61
28.75	80.05	87.53	89.15	51.47	57.31	61.21
34.50	78.46	87.10	89.04	51.57	58.02	61.57
40.25	77.11	86.67	88.92	51.66	58.99	61.72
46.00	75.76	86.25	88.82	51.69	60.13	61.81

5.5 Summary

This chapter analyzed the performance of MLGNR as a potential candidate to replace the Cu for future VLSI interconnects. Based on the multiconductor transmission line theory, the ESC model of MLGNR is presented. In a more realistic manner, the proposed model incorporates the width dependent MFP parameter that helps in accurately estimating the crosstalk induced performance in comparison to the conventional model. The proposed ESC model is validated by comparing its transient response with respect to the response of MTL model.

The FDTD model is presented to analyze the crosstalk effects in coupled MLGNR interconnect lines. The results of the proposed model closely match with that of HSPICE simulations. The average error in the propagation delay measurement is observed to be less than 2 %. Moreover, it has been noticed that the width dependent MFP should be incorporated in a valid crosstalk noise modeling. In addition, the efficiency of the proposed model has also been demonstrated. It is observed that the model requires at an average only 5 % of HSPICE simulation time. Based on the comparative study, it is observed that the MLGNR is the better suitable on-chip interconnect material than the Cu.

References

1. Rabaey JM, Chandrakasan A, Nikolic B (2003) Digital integrated circuits: a design perspective, 2nd edn. Prentice-Hall, New Jersey
2. Mezhiba AV, Friedman EG (2002) Inductive properties of high-performance power distribution grids. IEEE Trans Very Large Scale Integr (VLSI) Syst 10(6):762–776
3. Yi M, Swaminathan M, Qian Z, Aydiner A (2012) Skin effect modeling of interconnects using the Laguerre-FDTD scheme. In: Proceedings of 21st conference on electrical performance of electronic packaging and systems (EPEPS), Tempe, pp 236–239

4. Kumar VR, Kaushik BK, Patnaik A (2015) Crosstalk modeling with width dependent MFP in MLGNR interconnects using FDTD technique. In: Proceedings of IEEE conference on electron devices and solid-state circuits, Singapore, pp 138–141
5. Kumar VR, Kaushik BK, Patnaik A (2014) Modeling of crosstalk effects in coupled MLGNR interconnects based on FDTD method. In: Proceedings of IEEE electronic components and technology conference (ECTC), Florida, USA, pp 1091–1097
6. Naeemi A, Meindl JD (2008) Electron transport modeling for junctions of zigzag and armchair graphene nanoribbons (GNRs). IEEE Electron Device Lett 29(5):497–499
7. Stan MR, Unluer D, Ghosh A, Tseng F (2009) Graphene devices, interconnect and circuits— challenges and opportunities. In: Proceedings on IEEE international symposium on circuits and systems (ISCAS), Taipei, 24–27 May 2009, pp 69–72
8. Xu C, Li H, Banerjee K (2009) Modeling, analysis, and design of graphene nanoribbon interconnects. IEEE Trans Electron Devices 56(8):1567–1578
9. Li H, Xu C, Srivastava N, Banerjee K (2009) Carbon nanomaterials for next-generation interconnects and passives: physics, status and prospects. IEEE Trans Electron Devices 56(9):1799–1821
10. Kumar P, Singh A, Garg A, Sharma R (2013) Compact models for transient analysis of single-layer graphene nanoribbon interconnects. In: Proceedings of IEEE international conference on computer modelling and simulation (UKSim2013), Cambridge, 2013, pp 809–814
11. Xu C, Li H, Banerjee K (2008) Graphene nano-ribbon (GNR) interconnects: a genuine contender or a delusive dream?In: Proceedings of IEEE international electron devices meeting (IEDM 2008), San Francisco, CA, USA, 2008, pp 1–4
12. Moon JS, Gaskill DK (2011) Graphene: its fundamentals to future applications. IEEE Trans Microw Theory Tech 59(10):2702–2708
13. Zhao W-S, Yin W-Y (2014) Comparative study on multilayer graphene nanoribbon (MLGNR) interconnects. IEEE Trans Electromagn Compat 56(3):638–645
14. Cui J, Zhao W, Yin W, Hu J (2012) Signal transmission analysis of multilayer graphene nano-ribbon (MLGNR) interconnects. IEEE Trans Electromagn Compat 54(1):126–132
15. Avouris P (2010) Graphene: electronic and photonic properties and devices. Nano Lett 10(11):4285–4294
16. Berger C, Song Z, Li X, Wu X, Brown N, Naud C, Mayou D, Li T, Hass J, Marchenkov AN, Conrad EH, First PN, Heer WA (2006) Electronic confinement and coherence in patterned epitaxial graphene. Science 312(5777):1191–1196
17. Naeemi A, Meindl JD (2007) Conductance modeling for graphene nanoribbon (GNR) interconnects. IEEE Electron Device Lett 28(5):428–431
18. Nasiri SH, Faez R, Moravvej-Farshi MK (2012) Compact formulae for number of conduction channels in various types of grapheme nanoribbons at various temperatures. Mod Phys Lett B 26(1):1150004-1–115004-5
19. Sakurai T, Newton AR (1991) A simple MOSFET model for circuit analysis. IEEE Trans Electron Devices 38(4):887–894
20. Kumar VR, Kaushik BK, Patnaik A (2014) An accurate FDTD model for crosstalk analysis of CMOS-gate-driven coupled RLC interconnects. IEEE Trans Electromagn Compat 56(5):1185–1193
21. Zhang J, Friedman EG (2006) Crosstalk modeling for coupled *RLC* interconnects with application to shield insertion. IEEE Trans VLSI Syst 14(6):641–646
22. Bandyopadhyay T, Han KJ, Chung D, Chatterjee R, Swaminathan M, Tummala R (2011) Rigorous electrical modeling of through silicon vias (TSV) for MOS capacitance effects. IEEE Trans Compon Packag Manuf Technol 1(6):893–903

Chapter 6
An Efficient US-FDTD Model for Crosstalk Analysis of On-Chip Interconnects

Abstract This chapter introduces a novel unconditionally stable FDTD (US-FDTD) model for the performance analysis of on-chip interconnects. It is observed that the stability of the proposed US-FDTD model is not constrained by the CFL condition and is therefore unconditionally stable. The accuracy of the proposed model is validated against the conventional FDTD model. It is observed that the US-FDTD model is as accurate as the conventional FDTD model while being highly time efficient. Moreover, the performance of Cu interconnect is compared with MWCNT and MLGNR interconnects under the influence of crosstalk.

Keywords Courant–Friedrichs–Lewy (CFL) · Stability condition · Finite difference time domain (FDTD) · Unconditionally stable · VLSI interconnects

6.1 Introduction

The shrinking size of the transistors has resulted in gate delays being overshadowed by larger interconnect delays [1, 2]. Therefore, the overall chip performance is primarily dependent on the interconnect performance. The close proximity of interconnects in miniaturized microelectronic devices leads to crosstalk noise. The crosstalk noise may result in logic failure, circuit malfunction, change in signal propagation and unwanted power dissipation [3]. Therefore, accurate modeling of crosstalk noise has emerged as vital design criteria in microelectronics.

The FDTD technique is widely used to solve electromagnetic wave problems. It is a fast, accurate and robust technique, which involves discretization of electromagnetic fields in both space and time domains [4–6]. Recently, the application of this versatile technique has been extended to high-speed interconnect domain [7]. However, in the FDTD technique to ensure a stable operation, the time step size (Δt) is limited by the Courant–Friedrichs–Lewy (CFL) stability condition, i.e., $\Delta t \leq \Delta t_{max}$, where $\Delta t_{max} = \Delta z / v_{max}$, the terms Δz and v_{max} represent the space step size and the maximum phase velocity, respectively [8, 9]. Consequently, the

© The Author(s) 2016

B.K. Kaushik et al., *Crosstalk in Modern On-Chip Interconnects*,
SpringerBriefs in Applied Sciences and Technology,
DOI 10.1007/978-981-10-0800-9_6

conventional FDTD techniques [7, 10, 11] consume large memory space and power due to the enormous number of iterations required for the analysis. Hence, beyond the CFL condition, the FDTD technique is unstable and within it, the technique is not efficient.

The improvements in FDTD technique can be easily addressed if the CFL stability condition is removed. Recently, several researchers have proposed various modified FDTD techniques to overcome the CFL stability criteria based on different algorithms, such as alternating direction implicit (ADI)-FDTD [12, 13], split-step FDTD [14, 15], Crank-Nicolson (CN)-FDTD [16, 17] and others [18–20]. All these techniques [12–20] were developed for transmission lines that are usually excited and terminated by resistive drivers and resistive loads, respectively. However, the VLSI interconnects are driven and terminated by the nonlinear CMOS drivers and capacitive loads, respectively. Therefore, the existing unconditionally stable FDTD techniques are not suitable to analyze the performance of CMOS gate driven VLSI interconnects.

In this chapter, a novel model is proposed that successfully implements an unconditionally stable FDTD (US-FDTD) technique to analyze the comprehensive crosstalk effects of coupled VLSI interconnects. The interconnect lines are driven by the nonlinear CMOS driver that are modeled by the modified Alpha power law model, which includes the drain conductance parameter. The crosstalk induced performance parameters such as noise peak voltage, noise width, and delay are analyzed. The proposed model is compared against the conventional FDTD model for accuracy, efficiency and stability.

The remaining chapter is organized as follows: Sect. 6.2 details the implementation of the US-FDTD technique for coupled interconnect lines. Moreover, the unconditional stability of the model is also scrutinized. Section 6.3 analyzes the crosstalk noise and validates the accuracy and efficiency of the proposed model. Moreover, to verify the unconditional stability of the proposed model the transient analysis is carried out at different values of time step. The performance of Cu, MWCNT and MLGNR interconnects is compared in Sect. 6.4. Finally, Sect. 6.5 concludes the chapter.

6.2   Development of Proposed US-FDTD Model

This section deals with the development of US-FDTD model for the coupled on-chip interconnects. The interconnect lines are coupled capacitively and inductively. In a more realistic manner, the CMOS drivers are considered for accurate performance analysis. The interconnect lines are terminated by capacitive loads. The schematic view of coupled interconnect lines is shown in Fig. 6.1, where the line x and the line y represent the aggressor and victim lines, respectively.

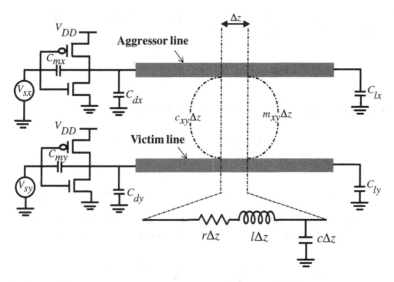

Fig. 6.1 Schematic view of coupled interconnects driven by CMOS drivers

6.2.1 Modeling of Coupled On-Chip Interconnects

The interconnect lines are represented by the transmission lines that are coupled capacitively, c_{xy} and inductively, m_{xy}. The telegrapher's equations in the transverse electromagnetic mode of coupled lines, at any point z along the line, is represented as

$$\frac{\partial v_x(z,t)}{\partial z} + r_x i_x(z,t) + l_x \frac{\partial i_x(z,t)}{\partial t} + m_{xy}\frac{\partial i_y(z,t)}{\partial t} = 0 \qquad (6.1\text{a})$$

$$\frac{\partial v_y(z,t)}{\partial z} + r_y i_y(z,t) + l_y \frac{\partial i_y(z,t)}{\partial t} + m_{xy}\frac{\partial i_x(z,t)}{\partial t} = 0 \qquad (6.1\text{b})$$

$$\frac{\partial i_x(z,t)}{\partial z} + c_x \frac{\partial v_x(z,t)}{\partial t} + c_{xy}\frac{\partial v_x(z,t)}{\partial t} - c_{xy}\frac{\partial v_y(z,t)}{\partial t} = 0 \qquad (6.1\text{c})$$

$$\frac{\partial i_y(z,t)}{\partial z} + c_y \frac{\partial v_y(z,t)}{\partial t} + c_{xy}\frac{\partial v_y(z,t)}{\partial t} - c_{xy}\frac{\partial v_x(z,t)}{\partial t} = 0 \qquad (6.1\text{d})$$

Representing Eqs. (6.1a) and (6.1b) in matrix form results in

$$\frac{\partial \mathbf{v}(z,t)}{\partial z} + \mathbf{r}\mathbf{i}(z,t) + \mathbf{l}\frac{\partial \mathbf{i}(z,t)}{\partial t} = \mathbf{0} \qquad (6.2\text{a})$$

where v and i are evaluated in 2×1. Matrix form as $v = [v_x, v_y]^T$, $i = [i_x, i_y]^T$, r and l are evaluated in 2×2 matrix form as

$$r = \begin{pmatrix} r_x & 0 \\ 0 & r_y \end{pmatrix}, \quad l = \begin{pmatrix} l_x & m_{xy} \\ m_{xy} & l_y \end{pmatrix}$$

Representing Eqs. (6.1c) and (6.1d) in matrix form results in

$$\frac{\partial i(z,t)}{\partial z} + c\frac{\partial v(z,t)}{\partial t} = 0 \qquad (6.2b)$$

where c is evaluated in 2×2 matrix form as $c = \begin{pmatrix} c_x + c_{xy} & -c_{xy} \\ -c_{xy} & c_y + c_{xy} \end{pmatrix}$.

Figure 6.2 represents the space discretization of an interconnect line. Using the forward difference approximation in space domain, Eq. (6.2a) results in

$$\frac{v_{k+1} - v_k}{\Delta z} = -ri_k - l\frac{\partial i_k}{\partial t} \qquad (6.3a)$$

For $k = 1, 2, \ldots, N_z$, Eq. (6.3a) can be rearranged as

$$v_{k+1} - v_k + r\Delta z i_k + l\Delta z\frac{\partial i_k}{\partial t} = 0 \qquad (6.3b)$$

Using the backward difference approximation in space domain, Eq. (6.2b) results in

$$\frac{i_k - i_{k-1}}{\Delta z} = -c\frac{\partial v_k}{\partial t} \qquad (6.4)$$

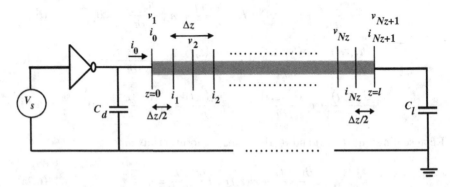

Fig. 6.2 Representation of space discretization of an interconnect line for unconditionally stable FDTD technique. The total interconnect length is divided into N_z number of sections, each with a uniform length of Δz

For $k = 1$ and $k = Nz + 1$, the space segment is replaced by $\Delta z/2$ (Fig. 6.2). For $k = 1$, Eq. (6.4) results in

$$i_1 - i_0 + \frac{c\Delta z}{2}\frac{\partial v_1}{\partial t} = 0 \tag{6.5a}$$

For $k = Nz + 1$, Eq. (6.4) results in

$$i_{Nz+1} - i_{Nz} + \frac{c\Delta z}{2}\frac{\partial v_{Nz+1}}{\partial t} = 0 \tag{6.5b}$$

For $k = 2, 3, \ldots, Nz$, Eq. (6.4) results in

$$i_k - i_{k-1} + c\Delta z\frac{\partial v_k}{\partial t} = 0 \tag{6.5c}$$

Equations (6.5a) and (6.5b) are further modified after applying the boundary conditions, as illustrated in the following Sects. 6.2.2 and 6.2.3.

6.2.2 Modeling of CMOS Driver

The CMOS drivers are modeled by modified alpha power law model that incorporates the effect of velocity saturation along with the finite drain conductance parameter. As shown in Fig. 6.1, the parasitic capacitances C_m and C_d represent gate to drain coupling capacitance and drain diffusion capacitance, respectively. The NMOS and PMOS transistors can operate in either cutoff, linear or saturation regions depending on the input voltage signal [21] as depicted in Fig. 6.3.

The current equations of the MOS transistors are

$$I_n = \begin{cases} 0 & \text{(cutoff)} \\ M_{Ln}(V_s - V_{Tn})^{\alpha_n/2}v_1 & \text{(lin)} \\ M_{Sn}(V_s - V_{Tn})^{\alpha_n}(U + \sigma_n v_1) & \text{(sat)} \end{cases} \tag{6.6}$$

$$I_p = \begin{cases} 0 & \text{(cutoff)} \\ M_{Lp}(V_{DD} - V_s - |V_{Tp}|)^{\alpha_p/2}(V_{DD} - v_1) & \text{(lin)} \\ M_{Sp}(V_{DD} - V_s - |V_{Tp}|)^{\alpha_p}(U + \sigma_p(V_{DD} - v_1)) & \text{(sat)} \end{cases} \tag{6.7}$$

where U is a 2×2 identity matrix, M_{Ln} (M_{Lp}), M_{Sn} (M_{Sp}), α_n (α_p), σ_n (σ_p) and V_{Tn} (V_{Tp}) are the linear region transconductance parameter, saturation region transconductance parameter, velocity saturation index, drain conductance parameter and the threshold voltage of NMOS (PMOS), respectively. The model parameters of NMOS/PMOS transistor at 32 nm technology node are listed in Table 6.1.

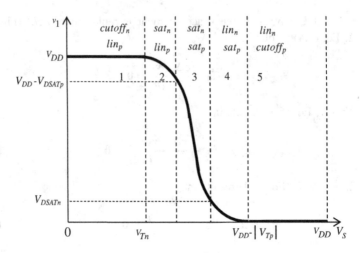

Fig. 6.3 The five regions of operation of a CMOS inverter. The subscripts n and p denote NMOS and PMOS, respectively

Table 6.1 Model parameters for 32 nm node

Parameter	PMOS	NMOS
M_L	0.006	0.007
M_S	0.875105×10^{-3}	1.944973×10^{-3}
α	1.0788	0.91503
σ	2.685	0.876
V_T	0.36	0.35

6.2.3   Modeling of Driver-Interconnect-Load

In this sub-section, the near-end and far-end interconnect terminal conditions are incorporated in the current equations. At the near-end terminal ($k = 1$), applying Kirchhoff's current law, the source current i_0 can be expressed as

$$i_0 = I_p - I_n + C_m \left(\frac{\mathrm{d}(V_s - v_1)}{\mathrm{d}t} \right) - C_d \frac{\mathrm{d}v_1}{\mathrm{d}t} \tag{6.8}$$

Incorporating Eq. (6.8) in Eq. (6.5a) results in

$$i_1 - I_p + I_n - C_m \left(\frac{\mathrm{d}(V_s - v_1)}{\mathrm{d}t} \right) + C_d \frac{\mathrm{d}v_1}{\mathrm{d}t} + \frac{c\Delta z}{2} \frac{\partial v_1}{\partial t} = 0 \tag{6.9}$$

At the far-end terminal ($k = Nz + 1$), the load current i_{Nz+1} can be expressed as

$$i_{Nz+1} = C_l \frac{\partial v_{Nz+1}}{\partial t} \qquad (6.10)$$

Incorporating Eq. (6.10) in (6.5b) results in

$$i_{Nz} - \left(\frac{c\Delta z}{2} + C_l\right)\frac{\partial v_{Nz+1}}{\partial t} = 0 \qquad (6.11)$$

Representing Eqs. (6.9), (6.5c) and (6.11) together in the matrix form results in

$$PI + GV + C\frac{\partial V}{\partial t} = I_s + X_r\frac{\partial V_r}{\partial t} \qquad (6.12)$$

where $P = \begin{pmatrix} U & 0 & \cdots & 0 & 0 \\ -U & U & \cdots & 0 & 0 \\ \vdots & \vdots & \ddots & \vdots & \vdots \\ 0 & 0 & \cdots & -U & U \\ 0 & 0 & \cdots & 0 & -U \end{pmatrix}_{2(Nz+1)\times 2Nz}$, $G = \begin{pmatrix} E_1 & 0 & \cdots & 0 & 0 \\ 0 & 0 & \ddots & \vdots & 0 \\ \vdots & \ddots & \ddots & 0 & 0 \\ 0 & \ddots & \ddots & 0 & \vdots \\ 0 & 0 & \cdots & 0 & 0 \end{pmatrix}_{2(Nz+1)\times 2(Nz+1)}$,

$C = \begin{pmatrix} \frac{c\Delta z}{2} + C_d + C_m & 0 & 0 & \cdots & 0 \\ 0 & c\Delta z & \ddots & \ddots & 0 \\ \vdots & 0 & \ddots & 0 & \vdots \\ 0 & \vdots & \ddots & c\Delta z & 0 \\ 0 & 0 & \cdots & 0 & \frac{c\Delta z}{2} + C_l \end{pmatrix}_{2(Nz+1)\times 2(Nz+1)}$, $I_s = \begin{pmatrix} E_2 \\ 0 \\ \vdots \\ \vdots \\ 0 \end{pmatrix}_{2(Nz+1)\times 1}$,

$I = \begin{pmatrix} i_1 \\ i_2 \\ \vdots \\ \vdots \\ i_{Nz} \end{pmatrix}_{2Nz\times 1}$, $V_r = \begin{pmatrix} V_s \\ 0 \\ \vdots \\ \vdots \\ 0 \end{pmatrix}_{2(Nz+1)\times 1}$, $V = \begin{pmatrix} v_1 \\ v_2 \\ \vdots \\ \vdots \\ v_{Nz+1} \end{pmatrix}_{2(Nz+1)\times 1}$,

$X_r = \begin{pmatrix} C_m & 0 & \cdots & \cdots & 0 \\ 0 & 0 & \ddots & \ddots & 0 \\ \vdots & \vdots & \ddots & \ddots & \vdots \\ \vdots & \vdots & \ddots & \ddots & 0 \\ 0 & 0 & \cdots & 0 & 0 \end{pmatrix}_{2(Nz+1)\times 2(Nz+1)}$

The values of E_1 and E_2 are dependent on the operating region of CMOS inverter and can be obtained from Table 6.2.

Representing Eq. (6.3b) in matrix form results in

$$QV + RI + L\frac{\partial I}{\partial t} = 0 \qquad (6.13)$$

Table 6.2 E_1 and E_2 for different regions of operation

Region	E_1	E_2
1	$M_{Lp}(V_{DD} - V_s - |V_{Tp}|)^{\alpha_p/2}$	$M_{Lp}(V_{DD} - V_s - |V_{Tp}|)^{\alpha_p/2}V_{DD}$
2	$M_{Lp}(V_{DD} - V_s - |V_{Tp}|)^{\alpha_p/2} + \sigma_n M_{Sn}(V_s - V_{Tn})^{\alpha_n}$	$M_{Lp}(V_{DD} - V_s - |V_{Tp}|)^{\alpha_p/2}V_{DD} - M_{Sn}(V_s - V_{Tn})^{\alpha_n}$
3	$\sigma_p M_{Sp}(V_{DD} - V_s - |V_{Tp}|)^{\alpha_p} + \sigma_n M_{Sn}(V_s - V_{Tn})^{\alpha_n}$	$M_{Sp}(V_{DD} - V_s - |V_{Tp}|)^{\alpha_p} + \sigma_p M_{Sp}(V_{DD} - V_s - |V_{Tp}|)^{\alpha_p}V_{DD} - M_{Sn}(V_s - V_{Tn})^{\alpha_n}$
4	$M_{Ln}(V_s - V_{Tn})^{\alpha_n/2} + \sigma_p M_{Sp}(V_{DD} - V_s - |V_{Tp}|)^{\alpha_p}$	$M_{Sp}(V_{DD} - V_s - |V_{Tp}|)^{\alpha_p} + \sigma_p M_{Sp}(V_{DD} - V_s - |V_{Tp}|)^{\alpha_p}V_{DD}$
5	$M_{Ln}(V_s - V_{Tn})^{\alpha_n/2}$	0

$$\text{where } Q = \begin{pmatrix} -U & U & 0 & 0 & \cdots & 0 \\ 0 & -U & U & \ddots & \ddots & 0 \\ \vdots & \ddots & \ddots & \ddots & 0 & \vdots \\ 0 & \ddots & \ddots & -U & U & 0 \\ 0 & 0 & \cdots & 0 & -U & U \end{pmatrix}_{2Nz \times 2(Nz+1)} , \quad R = \begin{pmatrix} r\Delta z & 0 & \cdots & 0 \\ 0 & r\Delta z & \ddots & \vdots \\ \vdots & \ddots & \ddots & 0 \\ 0 & \cdots & 0 & r\Delta z \end{pmatrix}_{2Nz \times 2Nz} ,$$

$$L = \begin{pmatrix} l\Delta z & 0 & \cdots & 0 \\ 0 & l\Delta z & \ddots & \vdots \\ \vdots & \ddots & \ddots & 0 \\ 0 & \cdots & 0 & l\Delta z \end{pmatrix}_{2Nz \times 2Nz}$$

Applying finite difference in time domain to Eqs. (6.12) and (6.13)

$$P\frac{I^{n+1}+I^n}{2} + G\frac{V^{n+1}+V^n}{2} + C\frac{V^{n+1}-V^n}{\Delta t} = \frac{I_s^{n+1}+I_s^n}{2} + X_r\frac{V_r^{n+1}-V_r^n}{\Delta t} \tag{6.14}$$

$$Q\frac{V^{n+1}+V^n}{2} + R\frac{I^{n+1}+I^n}{2} + L\frac{I^{n+1}-I^n}{\Delta t} = 0 \tag{6.15}$$

Solving Eqs. (6.14) and (6.15) results in

$$V^{n+1} = K_1\left(\frac{I_s^{n+1}+I_s^n}{2} + X_r\frac{V_r^{n+1}-V_r^n}{\Delta t} - K_2V^n - K_3I^n\right) \tag{6.16}$$

$$I^{n+1} = -K_4\left(K_5I^n + K_6\left(V^{n+1}+V^n\right)\right) \tag{6.17}$$

where $K_1 = \left(\left(\frac{G}{2}+\frac{C}{\Delta t}\right)-\frac{P}{4}K_4Q\right)^{-1}$, $K_2 = \left(\left(\frac{G}{2}-\frac{C}{\Delta t}\right)-\frac{P}{4}K_4Q\right)$, $K_3 = \left(\frac{P}{2}-\frac{P}{2}K_4K_5\right)$, $K_4 = \left(\frac{R}{2}+\frac{L}{\Delta t}\right)^{-1}$, $K_5 = \left(\frac{R}{2}-\frac{L}{\Delta t}\right)$, $K_6 = \frac{Q}{2}$.

Initially, the voltage and current values of interconnect line are set to zero. After exciting the input voltage source, the voltages along the line are evaluated for a specific time using (6.16) in terms of the previous values of voltage and current. Thereafter, the currents are evaluated using (6.17), in terms of these voltages and previous current values. The process is continued till the final time step is reached. This procedure can be demonstrated with the flowchart shown in Fig. 6.4.

From Eqs. (6.16) and (6.17), it can be observed that these equations are in implicit form and hence free from stability condition. Moreover, the conductance parameter (g) can be incorporated in the proposed model without affecting the unconditional stability criteria.

Fig. 6.4 Flowchart for evaluation of voltage and current using the proposed US-FDTD technique

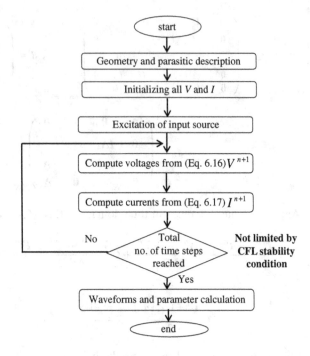

6.2.4 Stability Analysis

This subsection analytically demonstrates that the proposed US-FDTD model is not limited by the CFL stability condition and is unconditionally stable. Representing Eqs. (6.12) and (6.13) together in matrix form results in

$$AM + B\frac{\partial M}{\partial t} = \tilde{I}_s + \tilde{X}_r \frac{\partial \tilde{V}_r}{\partial t} \tag{6.18}$$

where $A = \begin{pmatrix} G & P \\ Q & R \end{pmatrix}$, $B = \begin{pmatrix} C & 0 \\ 0 & L \end{pmatrix}$, $\tilde{I}_s = \begin{pmatrix} I_s \\ 0 \end{pmatrix}$, $\tilde{X}_r = \begin{pmatrix} X_r \\ 0 \end{pmatrix}$, $\tilde{V}_r = \begin{pmatrix} V_r \\ 0 \end{pmatrix}$, $M = \begin{pmatrix} V \\ I \end{pmatrix}$. Performing finite difference in time domain, the Eq. (6.18) results in

$$A\frac{M^{n+1} + M^n}{2} + B\frac{M^{n+1} - M^n}{\Delta t} = \frac{\tilde{I}_s^{n+1} + \tilde{I}_s^n}{2} + \tilde{X}_r \frac{\tilde{V}_r^{n+1} - \tilde{V}_r^n}{\Delta t} \tag{6.19a}$$

After rearranging, Eq. (6.19a) results in

$$M^{n+1} = XM^n + \left(\frac{A}{2} + \frac{B}{\Delta t}\right)^{-1} \left(\frac{\tilde{I}_s^{n+1} + \tilde{I}_s^n}{2} + \tilde{X}_r \frac{\tilde{V}_r^{n+1} - \tilde{V}_r^n}{\Delta t}\right) \tag{6.19b}$$

where $X = \left(\frac{A}{2} + \frac{B}{\Delta t}\right)^{-1} \left(\frac{B}{\Delta t} - \frac{A}{2}\right)$.

Fig. 6.5 Comparison of spectral radius of amplification matrix between proposed US-FDTD and conventional FDTD model

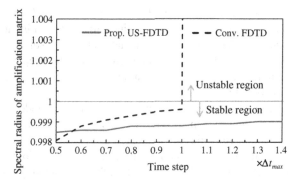

The matrix X is known as the amplification matrix of Eq. (6.19b). The model is stable if $\rho(X) \leq 1$, where $\rho(X)$ represents the spectral radius of X and can be expressed as [22]

$$\rho(X) = \max(|eigen(X)|) \tag{6.20}$$

With the increase in time step size, beyond the CFL limit, the conventional FDTD techniques [7, 10, 11] results in instability. However, for the proposed model the value of (6.20) is always less than 1 for all values of time step. Considering the worst case of the time step size, tending to infinity, the amplification matrix of the proposed US-FDTD model tends to identity matrix that has spectral radius of 1. Therefore, the proposed model is stable for any value of the time step size and hence, unconditionally stable.

The stability of the proposed model is analyzed based on the value of $\rho(X)$, which is calculated from Eq. (6.20). Figure 6.5 shows the comparison of $\rho(X)$ between the proposed US-FDTD model and conventional FDTD model, at different time steps. It is observed that when Δt is greater than the CFL limit (Δt_{max}), for the conventional FDTD model, $\rho(X) > 1$, whereas for the proposed US-FDTD model, $\rho(X) < 1$. Therefore, the proposed model is stable beyond the CFL stability condition as well [23]. The numerical example that demonstrates the unconditional stability of the proposed model is provided in Sect. 6.3.3.

6.3 Simulation Setup and Results

The proposed US-FDTD model is validated with the conventional FDTD model. Using the 32 nm technology, the width and thickness of the interconnect line are considered as 48 nm and 112.32 nm, respectively, with an aspect ratio of 2.34. The spacing between the interconnect lines and the height from the ground plane are assumed to be equal to the width and thickness, respectively. The interlevel metal insulator dielectric constant, load capacitance and length of the line are 2.78, 2 fF and 0.5 mm, respectively. The line capacitive and inductive parasitics are extracted from the electrostatic and magnetostatic field solvers, respectively [24, 25]. The symmetric CMOS drivers are used to drive the coupled interconnect lines [26]. The input to the aggressor line is a falling ramp signal from 0.9 to 0 V with a transition time of 50 ps.

The interconnect line supports two different modes of propagation, i.e., even and odd modes in coupled lines. For given line parasitics, the corresponding even and odd mode velocities are evaluated as $v_1 = 1.49 \times 10^8$ m/s and $v_2 = 0.49 \times 10^8$ m/s, respectively. To maintain high accuracy, the line discretization value (Δz) is considered as 0.5×10^{-5} m. Based on the CFL stability condition, the maximum time step size (Δt_{max}) while ensuring stable operation is 0.33×10^{-13} s.

6.3.1   Transient Analysis

The comprehensive crosstalk noise is analyzed at the far-end of the line 2 using proposed US-FDTD and conventional FDTD. The transient responses of functional, dynamic in-phase and dynamic out-phase crosstalk switching are shown in Figs. 6.6, 6.7 and 6.8, respectively. The input switching of interconnect lines is shown in the inset. For functional crosstalk, line 1 (aggressor) makes a transition from ground to V_{DD} while line 2 (victim) is at ground level. During dynamic in-phase crosstalk, both line 1 and line 2 switch from ground to V_{DD}. Finally, for dynamic out-phase crosstalk, line 1 and line 2 make the transition from ground to V_{DD} and from V_{DD} to ground, respectively. From Figs. 6.6, 6.7 and 6.8, it is observed that in all the three cases the proposed US-FDTD model accurately estimates the timing response with respect to the conventional FDTD model.

Fig. 6.6 Transient response comparison during the functional crosstalk

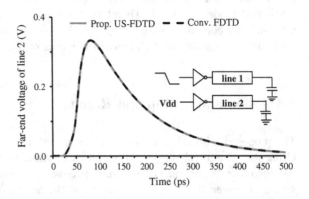

Fig. 6.7 Transient response comparison during the dynamic in-phase crosstalk

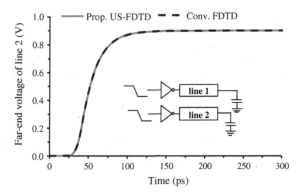

Fig. 6.8 Transient response comparison during the dynamic out-phase crosstalk

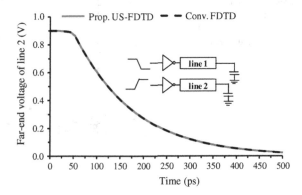

6.3.2 Crosstalk Induced Noise Peak, Width and Delay Analysis

When the generated noise has a peak voltage above the threshold voltage and attains a width above the threshold pulse width of the receiving gate, it leads to the generation of a glitch that may result in logical failure of the circuit [27]. Therefore, noise peak voltage and pulse width are the two important parameters for crosstalk noise analysis. Consequently, these parameters are estimated using the proposed US-FDTD model and conventional FDTD model. Table 6.3 shows the functional crosstalk noise peak voltage and pulse width at different load capacitances. The average percentage error is observed to be less than 1 % for noise peak voltage and noise width. It is also observed that as the load capacitance increases, the noise peak voltage decreases, whereas the noise width increases. This is because of the increase in time constant with the increase in load capacitance.

To further verify the accuracy of the proposed model, the dynamic in-phase and out-phase crosstalk induced delay are analyzed at different interconnect lengths. From Fig. 6.9, it can be observed that the proposed model accurately estimates the

Table 6.3 Noise peak voltage and noise width comparison between proposed US-FDTD and conventional FDTD simulations

Load capacitance (fF)	Peak voltage			Noise width		
	Proposed US-FDTD (V)	Conv. FDTD (V)	Error (%)	Proposed US-FDTD (ps)	Conv. FDTD (ps)	Error (%)
5	0.287	0.289	0.69	150.151	148.280	0.58
10	0.249	0.249	0	176.614	175.655	0.54
15	0.222	0.223	0.45	202.112	201.003	0.55
20	0.197	0.198	0.50	231.621	229.707	0.62

Fig. 6.9 Comparison of crosstalk induced propagation delay between proposed US-FDTD model and conventional FDTD model

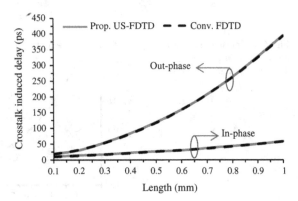

crosstalk induced propagation delay for all interconnect lengths, with an average error of less than 1 %. It is also observed that the out-phase delay is higher than the in-phase delay. This is due to the fact that the Miller capacitance strongly influences the signal propagation during the out-phase switching, whereas it is ineffective during the in-phase switching.

6.3.3 Unconditional Stability of the Proposed Model

The stability of the proposed model is verified through a numerical example demonstrating the dynamic in-phase transient response. The transient response is analyzed at different time steps using both conventional and proposed US-FDTD.

The stability of the proposed model is analyzed in the following two cases:

(1) Within the CFL condition ($\Delta t = \Delta t_{max}$), from Fig. 6.10a it can be observed that both conventional FDTD and proposed US-FDTD models provide stable outputs;

Fig. 6.10 Dynamic in-phase crosstalk analysis at different time steps **a** stable output of conventional FDTD and proposed US-FDTD and **b** unstable output of conventional FDTD due to the violation of CFL condition

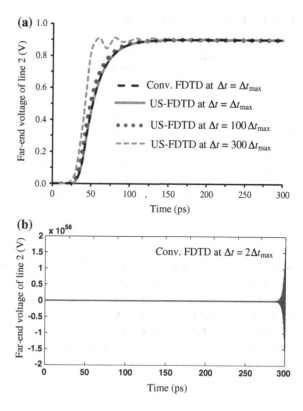

(2)   Beyond the CFL condition ($\Delta t > \Delta t_{max}$), from Fig. 6.10b it can be observed that the conventional FDTD model is unstable at $\Delta t = 2\Delta t_{max}$, whereas the proposed model is stable and sufficiently accurate, even when $\Delta t = 100 t_{max}$ (Fig. 6.10a). Although, at $\Delta t = 300\Delta t_{max}$ the proposed model output is stable but the accuracy reduces due to the lower resolution in temporal space. The stability of the proposed model is not limited by the CFL condition and is therefore unconditionally stable.

6.3.4   Efficiency of the Proposed Model

Using a PC with Intel Core i7 CPU (3.4 GHz, 8 GB RAM), the run-time of the proposed model is compared with the conventional FDTD. The computational effort for the transient analysis of dynamic in-phase crosstalk at different time steps is shown in Table 6.4. For $\Delta t \leq \Delta t_{max}$, the proposed US-FDTD model requires 89 %

Table 6.4 Comparison of computational efforts

Time step size, Δt	No. of iterations		CPU time (s)	
	Conventional FDTD model	Proposed US-FDTD model	Conventional FDTD model	Proposed US-FDTD model
$0.5\Delta t_{max}$	18,182	18,182	4.124	3.651
$1\Delta t_{max}$	9091	9091	2.056	1.824
$2\Delta t_{max}$	×	4545	×	0.914
$10\Delta t_{max}$	×	909	×	0.191
$50\Delta t_{max}$	×	182	×	0.037
$100\Delta t_{max}$	×	91	×	0.019

'×' represents the invalid instances since the conventional FDTD model becomes unstable for $\Delta t > \Delta t_{max}$

of the time utilized by the conventional FDTD model. It is worth noting that because the proposed model is unconditionally stable, it allows to consider larger time step, well beyond Δt_{max}. This further reduces the number of iterative steps required for analysis and consequently, reduces the computational run-time. As observed in Table 6.4, at $\Delta t = 100\Delta t_{max}$, where the conventional model results in unstable output, the proposed model requires only 0.9 % of the time required by conventional model at $\Delta t = \Delta t_{max}$. Therefore, depending on the time step size the proposed model can be up to 100 times faster than the conventional FDTD.

6.3.5   Comparison with 3D Simulations

The validity of the proposed model is further verified by the 3D simulations. Sentaurus TCAD simulator is used for the 3D simulations [28]. The interconnect width and spacing between the interconnect lines are equally considered as 48 nm, the thickness and height from the ground plane are equally considered as 112.32 nm. The interlevel metal insulator dielectric constant, load capacitance and length of the line are 2.78, 0.2 fF and 5 µm, respectively. The structure of the coupled interconnect line is shown in Fig. 6.11. The transient response at the far-end voltage of line 2 is observed during the functional crosstalk switching and shown in Fig. 6.12. It can be observed that the accuracy of the results via the proposed model is in good agreement with the 3D simulations. The average error between the proposed model and 3D simulation is less than 3 %. The CPU run-time and memory required by the proposed model are 0.02 s and 602 MB, respectively, whereas the 3D simulations require 964 s and 3066 MB, respectively.

Fig. 6.11 Structure of two coupled interconnects

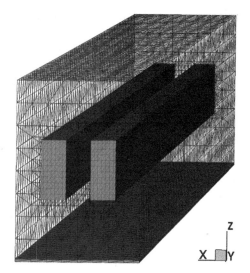

Fig. 6.12 Transient response comparison between the proposed US-FDTD model and 3D simulations at the far-end terminal of line 2 during the functional crosstalk switching

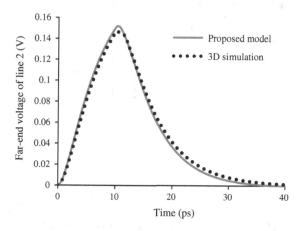

6.4 Performance Comparison of Cu, MWCNT and MLGNR Interconnects

In this sub-section, the propagation delay under the influence of crosstalk is analyzed among Cu, MWCNT and MLGNR interconnects. The simulations have been carried out using the US-FDTD method. The coupled interconnect line structures considered for the performance comparison of Cu/MWCNT/MLGNR is shown in Fig. 6.13. A CMOS driver with supply voltage $V_{DD} = 0.9$ V is used to drive the interconnect line. The relative permittivity of the inter layer dielectric medium is considered as 2.25. The width and spacing between coupled interconnects is equal to 48 nm and the distance from the ground plane is 86.4 nm.

Fig. 6.13 Structure of two coupled interconnects

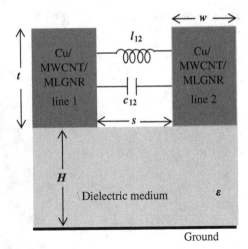

Fig. 6.14 Performance delay comparison among Cu, MWCNT and MLGNR interconnects

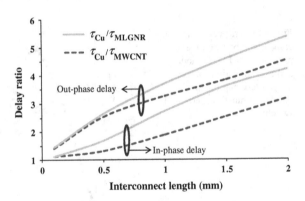

The crosstalk induced delay ratios of Cu to MLGNR (τ_{Cu}/τ_{MLGNR}) and MWCNT (τ_{Cu}/τ_{MWCNT}) are shown in Fig. 6.14. The AsF$_5$ doped MLGNR interconnect is used with intercalation doping density of $E_F = 0.4$ eV and back scattering probability of 0.2. The comparison plot shows that the crosstalk induced delay of MWCNT and MLGNR substantially reduces in comparison to the conventional Cu interconnect. This reduction is more prominent for out-phase switching due to the considerable reduction in coupling capacitance of MWCNT/MLGNR than Cu interconnects. Additionally, the crosstalk induced delay of MLGNR is smaller in comparison to MWCNT interconnects. The primary reason behind this reduced delay is the availability of more number of conduction channels in MLGNR than MWCNT. Therefore, the superior electrical properties and the fabrication compatibility with conventional lithography techniques make the MLGNR intrinsically more suitable material for on-chip interconnect realization. However, the intercalation doping of MLGNR with low back scattering probability is the primary challenge in designing a high performance MLGNR interconnect.

6.5 Summary

This chapter proposed a novel model based on US-FDTD technique, to analyze the crosstalk effects of coupled VLSI interconnects. It is analytically and numerically demonstrated that the proposed model is not limited by the CFL condition and is unconditionally stable. Using the proposed model, a comprehensive crosstalk analysis is performed and the results are in good agreement with the conventional FDTD model. The average percentage error is observed to be less than 1 % for the estimation of crosstalk induced performance parameters. Moreover, it is observed that the proposed model is highly time efficient than the conventional FDTD model. Depending on the time step size, the proposed model can be up to 100× faster than the conventional FDTD. Moreover, the performance comparison of Cu, MWCNT and MLGNR interconnects is also presented. It is observed that the MLGNR is the better suitable on-chip interconnect material than the Cu and MWCNT.

References

1. Zhang J, Friedman EG (2006) Crosstalk modeling for coupled *RLC* interconnects with application to shield insertion. IEEE Trans VLSI Syst 14(6):641–646
2. Koo KH (2000) Comparison study of future on-chip interconnects for high performance VLSI applications. Ph.D. thesis. Stanford University, United States
3. Das D, Rahaman H (2011) Analysis of crosstalk in single- and multiwall carbon nanotube interconnects and its impact on gate oxide reliability. IEEE Trans Nanotechnol 10(6):1362–1370
4. Livshits P, Sofer S (2012) Aggravated electromigration of copper interconnection lines in ULSI devices due to crosstalk noise. IEEE Trans Dev Mater Reliab 12(2):341–346
5. Jiang LL, Mao JF, Wu AL (2006) Simplistic finite-difference time domain method for Maxwell equations. IEEE Trans Magn 42(8):1991–1995
6. Bawa S, Sharma GK (2002) A parallel transitive closure computation algorithm for VLSI test generation. In: Proceedings of the 6th international conference on applied parallel computing advanced scientific computing, London, UK, pp 243–252
7. Farahat N, Raouf H, Mittra R (2002) Analysis of interconnect lines using the finite-difference time-domain (FDTD) method. Microw Opt Technol Lett 34(1):6–9
8. Taflove A (1995) Computational electrodynamics. Artech House, Norwood, MA
9. Strikwerda JC (1989) Finite difference schemes and partial differential equations. Brooks/Cole, Pacific Grove, CA
10. Liang F, Wang G, Lin H (2012) Modeling of crosstalk effects in multiwall carbon nanotube interconnects. IEEE Trans Electromagn Compat 54(1):133–139
11. Li XC, Ma JF, Swaminathan M (2011) Transient analysis of CMOS gate driven RLGC interconnects based on FDTD. IEEE Trans Comput Aided Des Int Circuit Syst 30(4):574–583
12. Lee YM, Chen C (2001) Power grid transient simulation in linear time based on transmission line-modeling alternating-direction implicit method. In: Proceedings of the IEEE international conference on computer aided design, ICCAD, USA, pp 75–80
13. Namiki T (1999) A new FDTD algorithm based on alternating-direction implicit method. IEEE Trans Microw Theory Techn 47(10):2003–2007
14. Wang W, Liu PG, Qin YJ (2013) An unconditional stable 1D-FDTD method for modeling transmission lines based on precise split-step scheme. Prog Electromagnet Res 135:245–260

15. Kong Y, Chu Q, Li R (2013) Two efficient unconditionally-stable four-stages split-step FDTD methods with low numerical dispersion. Prog Electromagnet Res B 48:1–22
16. Sun C, Trueman CW (2003) Unconditionally stable Crank-Nicolson scheme for solving two-dimensional Maxwell's equations. Electron Lett 39(7):595–597
17. Tan EL (2008) Efficient algorithms for Crank-Nicolson-based finite difference time-domain methods. IEEE Trans Microw Theory Tech 56(2):408–413
18. Tang M, Mao JF (2008) A precise time-step integration method for transient analysis of lossy nonuniform transmission lines. IEEE Trans Electromagnet Compat 50(1):166–174
19. Afrooz K, Abdipour A (2012) Efficient method for time-domain analysis of lossy nonuniform multiconductor transmission line driven by a modulated signal using FDTD technique. IEEE Trans Electromagnet Compat 54(2):482–494
20. Jia L, Shi W, Guo J (2008) Arbitrary-difference precise-integration method for the computation of electromagnetic transients in single-phase nonuniform transmission line. IEEE Trans Power Deliv 23:1488–1494
21. Kang SM, Leblebici Y (2003) CMOS digital integrated circuits: Analysis and design," 3rd edn. Tata McGraw-Hill
22. Kantartzis VN, Tsiboukis TD (2008) "Modern EMC analysis techniques volume I: Time-domain computational schemes. Synth. Lectures Comput. Electromagn., 1st edn., pp 1–224
23. Kumar VR, Alam A, Kaushik BK, Patnaik A (2015) An unconditionally stable FDTD model for crosstalk analysis of VLSI interconnects. IEEE Trans Compon Packag Manuf Technol 5(12):1810–1817
24. Maxwell 2D student version, Ansoft Corp., Pittsburgh, PA (2005)
25. FastHenry version 3.32 (2011) http://www.fastfieldsolvers.com
26. Tan MLP, Arora VK, Saad I, Ahmadi MT, Ismail R (2009) The drain velocity overshoot in an 80 nm metal-oxide-semiconductor field-effect transistor. J Appl Phys 105(7):074503-1–074503-7
27. Kahng AB, Muddu S, Vidhani D (1999) Noise and delay uncertainty studies for coupled RC interconnects. In: Proceedings of the IEEE international ASIC/SOC conference, Washington, DC, pp 3–8
28. Sentaurus user manuals (2015) Synopsys Inc. Mountain View, CA, USA

Printed in the United States
By Bookmasters